International Perspectives on Community Policing and Crime Prevention

Steven P. Lab

Criminal Justice Program
Bowling Green State University

Dilip K. Das

State University of New York at Plattsburgh

Upper Saddle River, New Jersey 07458

Library of Congress Cataloging-in-Publication Data

International perspectives on community policing and crime prevention / edited by Steven P. Lab, Dilip K. Das.
p. cm.
"Result of a three day conference sponsored by the International Police Executive Symposium that took place in The Hague in June 1998"—Pref.
Includes bibliographical references and index.
ISBN 0-13-030956-7
1. Community policing—Congresses. 2. Crime prevention—Congresses. I. Lab, Steven P. II. Das, Dilip K.

HV7936.C83 I56 2003
363.2'3—dc21 2001058000

Publisher: Jeff Johnston
Executive Editor: Kim Davies
Assistant Editor: Sarah Holle
Production Editor: Lori Dalberg, Carlisle Publishers Services
Production Liaison: Barbara Marttine Cappuccio
Director of Production & Manufacturing: Bruce Johnson
Managing Editor: Mary Carnis
Manufacturing Buyer: Cathleen Petersen
Design Director: Cheryl Asherman
Cover Design Coordinator: Miguel Ortiz
Cover Designer: Carey Davies
Marketing Manager: Jessica Pfaff
Composition: Carlisle Communications, Ltd.
Printing and Binding: Phoenix Book Tech Park

Pearson Education LTD.
Pearson Education Australia PTY, Limited
Pearson Education Singapore, Pte, Ltd.
Pearson Education North Asia Ltd.
Pearson Educatíon Canada, Ltd.
Pearson Educacíon de Mexico, S.A. de C.V.
Pearson Education—Japan
Pearson Education Malaysia, Pte. Ltd.

10 9 8 7 6 5 4 3 2 1
ISBN 0-13-030956-7

Contents

PREFACE

This book is the result of a three-day conference sponsored by the International Police Executive Symposium at The Hague in June 1998. The goal of the symposium was to bring together academics and practitioners from around the world to discuss community policing and crime prevention. Participants were asked to provide basic information about crime and policing in their respective countries, give a brief history of those efforts, and discuss the current state of preventive initiatives. What resulted was a very informative look at the diversity of policing, community policing, and crime prevention from around the world.

Those presentations, some of which developed into the chapters that appear in this book, attest to the diversity that can be found in community policing and crime prevention. Indeed, although the same terms may be used in different countries, the actual implementation can vary greatly. One may even argue that in some countries, such as China, community policing has been somewhat the norm throughout history and that many of the new initiatives are actually moving away from community participation. It is also possible to see that many countries face similar problems, but do so at different points in time, depending on larger social changes within the nation.

Everyone who participated in the symposium was invited to contribute a chapter for this book. Unfortunately, some individuals opted out. What remains however, is a fair sampling of the great diversity in community policing and crime prevention that can be found around the world. Would the book be better and more complete if the other presentations appeared here? Without a doubt! Nevertheless, a great deal of valuable information has been presented by those authors who eagerly participated.

I would like to thank my coeditor, Dilip Das, for identifying the meeting participants, coordinating the conference, and putting in the time and effort working with our hosts in The Hague. My gratitude is also extended to the following reviewers: Kenneth McCreedy, Ferrum College, Ferrum, VA; James Albrecht, NYPD, New York; Alex del Carmen, University of Texas, Arlington, TX; Will Oliver, Radford University, Radford, VA; and Stanley Swart, University of North Florida, Jacksonville, FL. Their input and expertise have greatly enhanced the scope and accuracy of this endeavor. I also want to thank and congratulate the authors of the individual chapters. They have all suffered through numerous rewrites, which posed a challenge for many of those for whom English is truly a foreign language. They are to be applauded for their patience as we worked to bring this book to final publication.

I sincerely hope that this information is of value to our readers.

Steven P. Lab

About the Editors

Steven P. Lab is Professor and Director of the Criminal Justice Program at Bowling Green State University. He is the author of *Crime Prevention: Approaches, Practices and Evaluations*, 4th ed. (Anderson Pub.), *Juvenile Justice*, 3rd ed. (with John Whitehead, Anderson Pub.) and *Victimology: An Introduction*, 3rd ed. (with William Doerner, Anderson Pub.), as well as numerous articles. Dr. Lab served as a Visiting Professor at Keele University in the United Kingdom during the time of the Fifth International Police Executive Symposium in The Hague.

Dilip K. Das is a professor at State University of New York (Plattsburgh). An author of several books and numerous articles, Dr. Das has been organizing annual symposia since 1994 as the Founder/President of the International Police Executive Symposium (IPES). He has also been a coauthor of the symposia-related books. He is the Editor-in-Chief of the journal, *Police Practice and Research: An International Journal*, which is an official publication of the IPES.

About the Authors

Alexis Aronowitz has conducted research on juvenile gangs for the Berlin government and has taught and served as Law Enforcement Program Manager for Central Texas College in Berlin and Kaiserslautern, Germany. She received the European Division Instructor of the Year Award in 1990. In 1994 she moved to The Netherlands to conduct research at the Research and Documentation Center at the Dutch Ministry of Justice. She has authored *Value-Added Tax Fraud in the European Union*, various book chapters, and articles in the areas of community policing, ethnic violence and hate crimes, and juvenile delinquency among ethnic minorities.

Obi N. I. Ebbe, an Ibo from eastern Nigeria, is a Professor of Criminal Justice at the University of Tennessee, Chattanooga. Dr. Ebbe is a recognized expert in the fields of political criminology and international criminal justice. Currently, he is an Associate Editor of Comparative and Applied Criminal Justice, a board member of *Trends in Organized Crime, International Journal of Police Science and Management*, and the African Editor (Research) for *Police Practice and Research: An International Journal.* He has published numerous articles in various academic journals. His latest work is *Comparative and International Criminal Justice Systems: Policing, Judiciary, and Corrections* (Butterworth/Heinemann Publishers, 1996).

Maximilian Edelbacher is a Hofrat (Chief Advisor) in the Federal Police of Vienna. Since 1988, he has been the Chief of the Major Crime Bureau in Austria. He graduated from the School of Law at the University of Vienna in 1968, and has been a legal expert in the Austrian Federal Police for over twenty-five years. He is a teacher in the Middle European Police Academy in Vienna and at the Vienna University of Economics and Business Administration. His publications include *Viennese Criminal Chronicle,* 1993, and *Applied Criminalistics* Series (3 volumes), 1995.

Ruth Geva was a Chief Superintendent and the Director of the Research and Development Section of the National Community Policing Unit at Police Headquarters in Israel. She has been involved in crime prevention in her country for the last twenty-five years. In 1973 she joined the Israel Police Service after completing a B.S. at the University of Toronto and an M.A. in Criminology, at the Faculty of Law at the Hebrew University in Jerusalem.

Yakov Gilinskiy is a Professor and the head of the Department of Sociology of Deviance (Center of Deviantology) of the Institute of Sociology of the Russian Academy of Sciences, Saint Petersburg Branch. He is also the Dean of the Department of Law at Saint Petersburg International Independent Baltic University of Ecology, Policy, and Law. Before becoming a University Scholar in 1968, Professor Gilinskiy was a barrister between 1958 and 1968. He has published extensively in the areas of criminology, sociology of deviance, and social control. He is a member of the Academy of Humanitarian Sciences and a member of New York Academy of Sciences.

Lloyd Hickman is a Superintendent with the Royal Canadian Mounted Police (RCMP). He served in the Protective Policing Directorate and the VIP Security Branch responsible for providing security to Canadian dignitaries, visiting foreign dignitaries, and foreign embassies. In 1983 he was assigned as the Personnel Security Officer with the Papal Visit Task Force and was in charge of the close personal protection of Pope John Paul II during his visit to Canada. Superintendent Hickman is currently the Officer in Charge of Lethbridge Sub-Division responsible for RCMP policing in southern Alberta. He attended the force's University program at Carleton University.

Peter C. Kratcoski is an Emeritus Professor of Criminal Justice Studies at Kent University. He served as Chairperson of the Department of Criminal Justice Studies at Kent State from 1986 to 1996. Currently, he is serving as an instructor and researcher in the department. His primary teaching and research responsibilities focus on corrections, sociology, juvenile delinquency, and victimology. Dr. Kratcoski is the Official Reporter of the International Police Executive Symposium.

R. J. Laing is a member of the Royal Canadian Mounted Police. Mr. Laing is actively involved in conducting research for the RCMP.

John Lindsay joined the Edmonton Police Service in 1976 with a B.A. degree from the University of Winnipeg. He completed a law degree at the University of Alberta in 1981 and was admitted to the Law Society of Alberta in 1982. Chief Lindsay served as the Edmonton Police Service Legal Advisor from 1982 until 1990 and was appointed a legal advisor to the Canadian Association of Chiefs of Police and the Standing Parliamentary Committee on Justice. Appointed as the Chief of Police in 1995, Chief Lindsay is a member of the Police Executive Research Forum, President of the Alberta Association of Chiefs of Police and a Vice-President with the Canadian Association of Chiefs of Police.

Mary M. Mwangangi is a Senior Assistant Commissioner of Police with the Kenya Police Service. Ms. Mwangangi is one of the very few women officers in the higher ranks of the service. She has been extensively involved in the new strategy of the Kenyan police to develop a community policing approach to crime prevention. She has had considerable experience in police-community relations and operations.

Miroslav Radovanovic, Ph.D., is a Professor of Sociology of Work and Poverty at the Faculty of Philosophy at the University of Belgrade. He is a member of various expert committees and councils on the general sociology and sociology of work and poverty as well as a member and Secretary of the editing board of *Sociology*. His special fields of interest are unemployment, poverty, working, and private freedom and rights. He has published more than 230 scientific papers on different topics in sociology.

R. K. Raghavan is a member of the Indian Police Service. Having served for many years in the Indian Intelligence Bureau with outstanding distinction, Dr. Raghavan is now the Director of the Central Bureau of Investigations.

A. Shiva Sankar is a member of the Indian Police Service holding the post of Inspector General of Police in the Indian State of Andhra. The Indian Police Service officers work for the national government as well as in all states in the Republic of India.

Irene Sárközi is a Major with Hungarian Police. She attended Teacher Training College from 1976 to 1979. She was a librarian in the Culture House of the Ministry of the Interior (1971–1981) and the Police Training School from 1981 to 1984. Ms. Sárközi was a student of Police College from 1984 to 1987. She served as an educational expert in the Ministry of the Interior and a researcher in the Police Research Institute, and was posted to the Janus Pannonius University, Department of Human Resources Management, from 1996 to 1998. At the present, Ms. Sárközi is the

Project Manager of Community Policing Program at the Hungarian National Police Headquarters. Her publications include *Crime in Hungary* and *Domestic Violence: Hungarian Experience.*

Rick Sarre is an Associate Professor at the School of International Business in the University of South Australia. He holds an LL.B. and a master's degree in Criminology. He has written extensively on policing, victimology, gun control, deviance, and private security. Sarre is the author of *Uncertainties and Possibilities: A Discussion of Selected Criminal Justice Issues in Australia* (1994), and has edited collections of articles on *International Victimology* (1996), *Sentencing and Indigenous Peoples* (1998), and *Exploring Criminal Justice: Contemporary Australian Themes* (1999).

Branislav Simonovic is a Professor of Criminalistics at the University of Kragujevac as well as at the University of Belgrade. He has published numerous scientific papers and a book, *Providing and Assessing a Statement in Police and Court.*

Walter Beller Taboado was a Professor of Philosophy (1982–1993) and later a Professor of Political and Social Science (1987–1992) in the Faculty of Philosophy and Literature at UNAM, the premier national university in Mexico. At present, he is the Director General of Crime Prevention and Community Services in the Office of the Attorney General of the Republic of Mexico. He is an author of many books, book chapters, and articles in the Spanish language.

Kam C. Wong is an Assistant Professor of Law at the Department of Government and Public Administration, Chinese University of Hong Kong, where he specializes in public law, criminal justice, and sociology of law. Professor Wong was an Inspector of Police with the Hong Kong Police and was awarded the Commissioner's High Commendation. Recently, he was awarded the Remington Award for distinguished work in law and social science by the School of Criminal Justice, SUNY (Albany). Professor Wong's publications appeared in *Journal of Asian Law, Occasional Papers in Contemporary Asian Studies, China Review* and *International Journal of the Sociology of Law.*

Introduction: Community Policing and Crime Prevention

Steven P. Lab

Any attempt to list recent initiatives aimed at improving the crime preventive capacities of the police invariably places community policing at, or near, the top of the list. Indeed, in the United States and most of the western world, community policing has emerged over the past ten to fifteen years as a major thrust of police activity. At the same time, the idea of community policing (or at least its rhetoric) has begun to take hold in countries across the world—from emerging democracies to highly regimented states. What immediately becomes apparent, however, is that what constitutes community policing varies from country to country, and what is viewed as new or innovative also differs across the globe.

Community policing should be viewed as an extension of attempts to build cooperative relations between the police and members of the public. In virtually every country there are programs and initiatives that seek out citizen input and assistance in dealing with crime and other social ills. Programs such as neighborhood watch, education programs aimed at young people, physical security improvements for homes, and targeted police patrols appear in many countries under many different headings other than "community policing." At the same time, most crime prevention programs rely on the input and support of police personnel. The difference between crime prevention and the "new" community policing is often little more than a change in the name and the explicit inclusion of the police in the process. The goals and techniques of community policing may be substantially the same as past prevention initiatives.

This introduction seeks to accomplish several objectives. First, if we are to discuss community policing, we must try to define the term. Although a single definition may be elusive, several common features can be delineated. Second, we need to answer the question: Is community policing new, or is it a repackaging of past preventive initiatives and practices? A key part of this introduction attempts to show that community policing is not really a new idea. Rather, it grows out of a long history of citizen participation in prevention activities, whether those be formal or informal. Finally, this section offers a brief introduction to all the chapters in this volume. The overall intent is to demonstrate the universal aspects of community policing and crime prevention and to offer insight into community policing based on experiences around the world.

WHAT IS COMMUNITY POLICING?

Trying to define community policing is like trying to hold mercury in your hand: It never looks exactly the same twice, and it is almost impossible to predict what will happen next. Indeed, there is no single definition of community policing that has been embraced by all writers or practitioners. Even the term "community policing" has variations, including the common titles of "community-oriented policing" and "problem-oriented policing." Interestingly, rather than delineate a single definition of what constitutes community policing, most discussions define it either by pointing out what it is *not* or by discussing the common elements that are typical of community policing initiatives.

Many police agencies have developed community policing into a process that is little more than police-community relations. Agencies have recognized the need to enhance interaction between officers and the citizenry in order to increase the level of citizen cooperation with the police. Specifically, the police need the citizens to report crime and provide information for prosecutions. The failure to engender such citizen participation inevitably results in low crime clearance rates and artificially low official crime statistics. The lack of interaction also results in distrust and suspicion on the part of both the public and the police toward each other. Many communities have taken steps to increase the "noncrime" interaction between citizens and the police. *Noncrime interaction* refers to instances where the two parties can meet and talk at a time when no crime has been committed, no one has been victimized, nobody needs assistance, or no specific outcome is anticipated. It is a means of letting the police and the public "get to know one another" outside of a confrontational or service situation. Although this activity should be encouraged, this is *not* community policing; it is rather, community relations.

Other police agencies have moved in the direction of identifying and attacking specific problems in the community. These agencies typically use some variation of Goldstein's (1990) problem-oriented policing approach. In essence, these agencies recognize that simply responding to calls for assistance, writing reports, making arrests, and gathering evidence for prosecutions is not enough. There is a need to be more proactive. Officers should try to identify problem situations before crime occurs and to propose potential solutions to problems. These solutions may involve traditional police activities, such as patrol and arrest, or may seek unique new law enforcement responses. Each situation may offer issues and problems not seen in the past or elsewhere. The solution, therefore, needs to be tailored to the problem. Solving the basic problem may eliminate the need for later coercive police activity. Problem-oriented policing, however, is *not* the same as community policing. The ideas are similar, but they are not identical. A key difference is that problem-oriented policing does not necessarily rely on or include citizens or agencies beyond the police.

So what is "community policing"? Most advocates describe it as a new approach to policing that embraces working with community members to address specific problems and the underlying causes of crime and deviance. Key elements of community policing typically include decentralizing the police through some type of community-based office or patrol, actively working to include citizens and other groups or agencies in the identification of problems, relying on a problem-solving approach that embraces the use of diverse legal and extralegal interventions, and establishing benchmarks (other than arrests and crime reduction) for assessing police activity. The underlying idea is that the police are not capable of dealing with the root causes of crime on their own. Instead, the police should become "community managers" (Hoover, 1992) who take a leadership position in identifying problems, building coalitions of individuals and groups appropriate for attacking the problems, overseeing the implementation of solutions, and making certain that evaluations of interventions are undertaken. The police need not be involved in the actual intervention or assessment. Rather, they should simply supervise these activities. Individuals, groups, or agencies who have the appropriate expertise or resources should undertake the intervention.

Community policing, therefore, should be seen as a philosophy of policing, rather than as a specific set of activities or approaches. This does not mean that community policing should be viewed as some amorphous idea. Indeed, it will be easier to embrace and promote the idea if it can be given some form or substance. One way to do this may be to relate these "new" ideas to past practices.

IS COMMUNITY POLICING NEW OR UNIQUE?

The basic ideas of community policing can be found in any discussion of early forms of social control. Until the establishment of formal policing in the 1800s, the primary responsibility for providing protection (either for the individual or the community) resided with the citizenry. Many early legal codes outlined the fact that the individual (or his family) was responsible for addressing transgressions against himself. The Code of Hammurabi and the Justinian Code both provided for retribution by the victim or his family.

One early form of citizen policing appeared in England after the Norman Conquest in 1066 (Klockars, 1985). Over the next two centuries, the idea of obligatory participation in actions to protect citizens and communities became the norm. All male citizens were required to participate in "policing" by responding to calls for help, assisting in the apprehension of offenders, and otherwise enforcing laws. In the early United States, vigilante groups were formed when needed to apprehend offenders and impose punishments. In essence, the community was responsible for protecting itself and responding to threats.

The advent of formal, paid police agencies and officers signaled a move away from reliance on community members to provide for their own protection and prevention. The citizenry turned over responsibility for combating crime to the police when large, professional agencies evolved in many countries. This movement, however, is relatively recent, dating from the early 1800s in westernized countries and within the last century in many communist, postcommunist and emerging third world countries. The move toward formal policing systems was accompanied by the growth of the scientific study of crime and deviance, and the belief that trained experts were needed to apprehend and deal with offenders. The police became the experts and the protectors of society. Members of the public became clients to be protected, sources of information, or offenders to be apprehended.

The growth of formal policing has greatly reduced the role of the citizenry in law enforcement. The police, however, still turn to the citizenry for assistance when the need arises. Indeed, elements of community policing appear throughout many police initiatives in many different countries. Foot patrol and neighborhood police offices are two examples of efforts to bring the police into greater contact with the citizenry. The expressed intent of these programs is to build cooperation between the police and the citizens in order to identify community problems and potential solutions. Another prime example of involving citizens is the emphasis on neighborhood watch and other crime prevention activities. The United States, the United Kingdom, and many countries have seen community crime prevention flourish in recent years. Even apparently traditional police activity, such as targeting specific problems or problem locations for intensive enforcement efforts, qualify as a move back to community involvement.

Beyond these apparently new efforts at community involvement, there is evidence that the basic proposition of community involvement and responsibility for dealing with deviance has never disappeared in many places. In many African and East Asian countries the citizens have been instrumental in maintaining order and responding to transgressions in local communities. The head of the household has retained responsibility for controlling the behavior of household members and taking action when an outsider harms the family. Formal policing efforts have been restricted largely to major metropolitan centers, and often concentrate on the protection of businesses, industry, the government, and the most affluent members of society. It can be argued that the move toward formal, professional police agencies is counter to a strong *community* policing philosophy, which has existed in some countries all along. Any discussion of building and incorporating community policing into formal police operations of some countries is almost counterproductive to the history of social control in those locations. In essence, community policing has always been the norm in many places.

The question of whether community policing is new or unique can be answered with a simple no. The underpinnings of community policing have their roots in the historical fact that it was the individual, his family, or the immediate community that were responsible for dealing with crime and deviance throughout most of history. Only within the last two centuries have formal social control agencies (particularly the police) assumed this role. Indeed, the ideas of individual or communal responsibility persist in many cultures and societies. In some sense, formal policing is returning to its roots when it looks to involve the larger community in its activities.

COMMUNITY POLICING AND CRIME PREVENTION

Crime prevention practitioners and those who study crime prevention initiatives can point out strong similarities between community policing and crime prevention. Indeed, it is possible to view community policing as the most prominent contemporary crime prevention approach. Many precursors to community policing are found in past crime prevention activities. As noted above, neighborhood or block watch programs have been the basis for drawing the public into active cooperation with the police. Local residents have been asked to join with the police for the purposes of identifying problems, taking precautions, and offering suggestions for eliminating the causes of problems. Community policing seeks to do essentially the same thing.

Perhaps the most obvious ties between community policing and crime prevention are embodied in the idea of *situational crime prevention*. Clarke (1983) defined situational prevention as taking measures to alter those factors of the immediate physical environment that provide the opportunity for specific forms of crime to occur. He proposed changes aimed primarily at the physical environment, but the basic idea is easily expanded to include the social environment. Underlying situational prevention is the premise that solutions to crime and deviance must assume diverse forms based on the uniqueness of the behavior and the context within which the behavior takes place. The same behavior committed in two different locations at two different times may require totally different responses. What is required, therefore, is a problem-solving approach, not unlike that implicit in community policing.

Both situational crime prevention and community policing share many of the same ideas and techniques. The SARA (*S*canning, *A*nalysis, *R*esponse, and *A*ssessment) process (Eck and Spelman, 1987) appears in both discussions of situational prevention and community policing. Although the terminology may shift, the basic idea is that a problem needs to be studied in order to identify potential solutions, and any implemented solution must be evaluated to assess its effectiveness. The problem dictates which individuals and agencies will be involved in the solution. Perhaps the only significant difference is that community policing assumes that the police are an integral part of the process, whereas situational prevention may or may not require police involvement.

The fact that community policing seeks to identify and ameliorate the underlying causes of crime clearly places it within the realm of crime prevention. Rather than simply wait to respond to calls, undertake investigations, make arrests and assist in prosecutions, community policing seeks to take action and *prevent* behavior *before* a crime has occurred. Community policing is crime prevention.

SCOPE OF THE BOOK

The chapters in this book offer insight to community policing and crime prevention as they appear in a variety of countries. All the chapters are based on presentations made at the Fifth International Police Executive Symposium, held at The Hague, The Netherlands, in June 1998. The authors have attempted to outline major community policing and crime prevention initiatives in their respective homelands. What becomes immediately apparent is both the diversity of the programs and the similarity in approaches.

The level of diversity across countries largely reflects the uniqueness of the crime problem and the history of social control in different countries. For example, some countries have a history of strong familial and informal community control over individuals. China has traditionally relied on various forms of informal methods to control deviants and prevent offending, particularly in rural areas. The relatively recent growth of formal policing has yet to take a firm hold in many areas, and true policing by the "community" remains an important part of social control. At the other end of the spectrum are countries like the United States with well-established police agencies that have all but eliminated informal modes for dealing with crime. In these countries, community policing and citizen crime prevention are "new" approaches, which must overcome the accepted, formalized social control systems.

Even though a great deal of diversity is present, it is interesting to note the similarities that exist. Many common themes appear throughout the chapters. Concern over engendering citizen cooperation and a lack of acceptance of the police by the public are almost universal issues. Similarly, attempts to increase informal, non-crime-related interaction between the police and the public are prominent in nearly every country. Activities that involve young people, businesses, community leaders, and family members are common. In almost every case, there is an emphasis on developing responses that target the specific problems and needs of local communities, rather than simply borrowing programs from other locales, which may or may not work in another area. Perhaps the greatest indication of shared concern and a desire to exchange ideas is the eager attendance and participation of everyone at the symposium.

ORGANIZATION OF THE BOOK

The various contributions to the book are loosely organized into several groups. The first set of chapters deals with countries that have well-established, formal, professional police forces. These policing organizations range from highly centralized structures to locally controlled operations. Most appear in North America and Europe. Included here are discussions of policing and crime prevention in Canada, The Netherlands, Australia, the United States, and Austria. In many of these countries, the recent emphasis on professionalism has led to situations where the police and public view crime and deviance as the province of formal social control agencies. The public has been removed as a truly active participant. In these countries the police are in the position of rebuilding strong ties with the citizenry and relying on the community for assistance in dealing with crime.

Several chapters come from countries that are relatively young or in transition from colonial status to independence. Among these countries are Israel, India, Kenya, and Nigeria. All these nations have been influenced by western forms of criminal justice, particularly through their past experiences as a part of the British empire. The basis of social control often includes both informal familial or communal actions taken primarily in rural areas and the use of formal police actions in urban centers. There is also a major tendency for the police to target crimes against businesses and the government in these countries, while neglecting offenses to individuals.

Similarly, three presentations offer insight to policing in emerging democracies. Yugoslavia, Hungary, and Russia are all former members of the Soviet Union and are currently facing the demands and trials associated with major political and social changes. The challenges facing the police, the criminal justice system, and crime prevention in these countries are quite different from those found in many other places. In the past, social control was the province of strong centralized authorities tied closely with the political regime, often involving military forces. Today, the economic conditions of the countries contribute to growing social problems and limit the potential responses. Involving the citizenry in preventive activities may offer some relief to overburdened social agencies.

Two additional chapters examine community policing and crime prevention in Mexico and China. Although both countries can point to strong centralized governmental control over the police, they exist within very diverse social settings. Mexico's policing and crime prevention initiatives can be found in various national plans and programs. Primary efforts target organized crime and drug trafficking. Concurrent with strong police involvement, there is an emphasis on family influence, education, publicity campaigns, and the promotion of values and ethics. For China, community policing and crime prevention can be viewed as a simple extension of age-old social practices. The family and local community have maintained primacy in dealing with deviant behavior, and the central government continues to promote such activities over the use of more formalized agency intervention. At the same time, there is evidence that more formal methods of control (including the police) are becoming increasingly necessary as the country is influenced by outside forces.

The final chapter offers a look at the future of community policing and crime prevention. Kratcoski, Das, and Verma try to draw out the major themes from the various chapters and provide direction for future practitioners and researchers. Despite the diversity in social climates and situations across countries, a number of recurring ideas and issues appear. It is the lessons to be learned from this information that form the basis for this final chapter.

Taken as a whole, the chapters offer insight into the current state of community policing and crime prevention in countries around the world. The chapters are not meant to provide a complete list of ideas or programs available for building cooperation between formal social control agencies, the police, and the general public. Instead, the programs and discussions offered by the authors show that there are many diverse approaches to problems of crime and deviance. At the same time, there is a unifying belief that the public is often a rich, untapped source of ideas and assistance for the police and other criminal justice agencies.

SECTION 1

The chapters in this book have been grouped into several sections based on general similarities between the countries. For example, the countries described in this first section have well established, formal, professional police forces. All these nations have a relatively long history of stability in their policing and legal systems. This does not mean that they have stagnated and are not undergoing change. Rather, the change tends to reflect the move toward professionalism and responses to emerging law enforcement concerns. This professionalism has often removed the public from active participation in policing and crime prevention, leaving the formal agents of social control to deal with crime and deviance. More recently, however, the police find themselves in the position of rebuilding or establishing new ties to the communities they serve. The chapters in this section reveal that this is a common challenge in each of these countries.

At the same time, the police forces and law enforcement efforts in these countries are not identical. One of the most notable differences is the degree of centralization in each nation. In the United States, for example, policing is mainly localized, with each community having a great deal of input into the size, shape, and operations of its police force. Much of this is due to the local funding of law enforcement in the country. More nationalized policing appears in Austria and Canada, where the central government plays a much larger role in the funding and operations of law enforcement. The local communities have less direct influence over the police force. One goal of the chapters is to educate the reader about how policing works in the different countries. The forms of community policing and crime prevention in these countries are perhaps more recognizable to many readers.

This organization of chapters is somewhat arbitrary, and some readers may feel that a different grouping would be preferable. We do not dispute that certain alternatives would have some value. But, each of the possible organizations has its merits. We hope that the one we have chosen makes good sense.

Chapter 1

Community Policing as Crime Prevention in the United States

Steven P. Lab

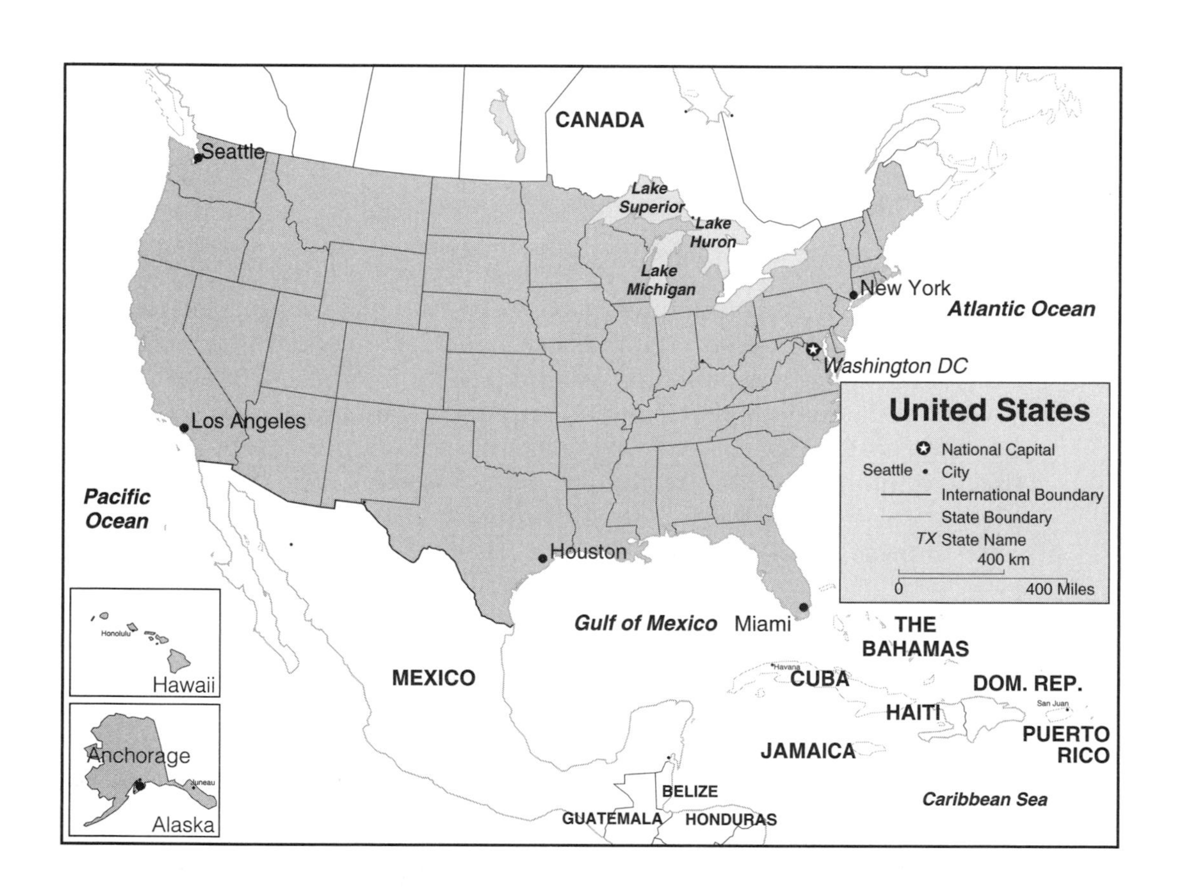

In recent years, the search for the most effective law enforcement in the United States has embraced the idea of community policing. Any inspection of textbooks, journals devoted to policing or law enforcement, governmental directives and policies on policing, or other similar sources uncovers almost unanimous agreement that community policing is the most appropriate direction for the future. This move is a result of the failure of traditional law enforcement methods to adequately eliminate crime and the fear of crime, and the recognition by law enforcement agencies that they cannot solve social problems on their own.

Although community policing is a driving force behind much police activity, many unanswered questions surround this approach. Foremost among these is the matter of defining "community policing." Related to the definitional problem is the issue of where community policing comes from and what form it should take. A final general area of concern involves a range of problems related to preparing both the police and the public to accept and implement community policing. This chapter attempts to address these issues as they have evolved in the United States.

DEFINING COMMUNITY POLICING

A quick examination of materials written on community policing reveals that no single, accepted definition exists. Instead, police departments, policy makers, researchers, and others use a wide array of definitions. Thankfully, most definitions revolve around a set of common themes. These themes outline a basic philosophical approach to the ways in which the police should attack crime and disorder in the community. At the same time, the definition of community policing in the United States often reflects simple participation in the federal Community-Oriented Policing program. Each of these is dealt with below.

A Philosophical Definition

As noted above, no single definition has emerged for community policing. One definition notes that

> "Community Policing" represents a fundamental change in the basic role of the police officer, including changes in his or her *skills, motivations,* and *opportunity* to engage in problem-solving activities and to develop new partnerships with key elements of the community (Wilkinson and Rosenbaum, 1994:110).

Similarly, Weisel and Eck (1994:51) define community policing as

> A diverse set of practices united by the general idea that the police and the public need to become better partners in order to control crime, disorder, and a host of other problems.

Many authors offer definitions that also deal with problem solving, partnerships, new approaches, and concern for more than just crime in the community.

Four key themes permeate the various definitions, and it may be more appropriate to consider community policing in terms of those themes or elements rather than to search for a single definition. In essence, it may be easier to identify common philosophical elements than it is to offer a definition on which everyone will agree. The essential elements or features of community policing appear to include community involvement, problem solving, a community base, and redefined goals.

Clearly, community policing requires cooperation between the police and varied community constituencies, including individual citizens, citizen groups, business associations, other local agencies (such as health departments, building inspectors, and community development offices), legislative bodies, and many others. Each group brings different ideas, abilities, and perspectives to the discussion of crime, fear, and community problems. Not all individuals or groups will opt to participate in every problem. Indeed, at times certain groups may not be invited to participate, often because the issue falls outside their purview or abilities.

Once the groups are brought together, community policing emphasizes *problem solving.* This is perhaps the most important element of community policing and originates from the premise that crime is a symptom of more basic concerns or problems in society. Whereas traditional policing attacks the symptom (a crime) through investigation, arrest, and assisting with prosecution, community policing shifts the emphasis to identifying and attacking the problems that underly or cause the crime to occur. Traditional police practices may eliminate some crime through deterrence or incapacitation, but they cannot impact problems not covered by the criminal code. Under community policing, the criminal code and legal action represents only one avenue for dealing with social problems. Community policing should approach issues and problems according to the uniqueness of each situation. Different problems may require vastly different solutions or interventions. The police, therefore, must identify and pursue solutions to the root problem.

A critical third component of the community policing philosophy is the decentralization of police operations. A centralized police organization, with officers working out of one or a few offices, patrolling in automobiles, and interacting with the public only in response to calls for assistance, is inappropriate for true problem solving. The police need to become an integral part of the community if they want to understand the underlying problems and issues facing the residents. Simply passing through an area on patrol or responding to calls about crime (a symptom of a larger problem) is not enough. Building a true community base can be accomplished through neighborhood stations or storefront offices, undertaking foot or bicycle patrol, assigning officers to a set shift and area over a long time period, and involving the public in setting police policies. The assumption is that officers will get to know the community, its problems, and its citizens.

The final major component to the community policing philosophy involves altering the goals of policing. Traditionally, the major goals have focused on making arrests and increasing the clearance rate for crime, and the public typically judges the police in terms of these goals. These goals, however, deal only with the overt symptoms of underlying problems in the community. Community policing seeks to shift the attention to the root causes of crime and other social problems. Community policing initiatives, to the extent that they deal with underlying causes and involve officers in nonarrest activities, require that the department and officers be held to different standards. Goals may include the reduction of crime, increased feelings of safety, the elimination of problem properties, less neighborhood disorder, increased community cohesion, improved grades among local youths, the establishment of new businesses in the community, and improved landscaping, among others. However, the important point is that community policing should emphasize the *ends,* rather than the *means* to the ends. Rather than focus on how things get done, the primary concern is the elimination of the problem. Successfully addressing root problems should eliminate the overt symptoms (that is, crime and fear of crime).

These four key ideas appear in most discussions of community policing. Other factors that appear in the literature include establishing more flexible organizational structures, focusing on disorder, altering the training for officers, and mobilizing citizens and other agencies (Carter, 1995; Eck and Rosenbaum, 1994; Hope, 1994; Wilkinson and Rosenbaum, 1994). Each of these fit under the four major themes. Most important is that all these ideas must work in unison. Just having one component, such as community offices, is not enough. Community policing requires making fundamental changes in philosophy, strategy, and programming (Cordner, 1995).

A Policy Definition of Community Policing

Although key philosophical aspects of community policing can be delineated, the reality of much community policing in the United States reflects the policy decisions of the federal government, which may or may not correspond to the philosophy. The federal government established the Community-Oriented Policing Office (COPS) in 1994. The major goal of this office was to fund local agencies in the hiring of 100,000 new "community police officers." A large part of the growth of community policing in the United States can be attributed to the funding initiatives of this office.

Gauging the success of community policing in terms of the simple hiring of "community police" officers suggests that the approach is well established in the United States. A closer inspection of what these new officers do, however, leads to a more conservative assessment of the growth and implementation of community policing. First, the COPS office does not delineate exactly what community policing means or how the new officers are to be used. The only expectation is that the community will use the new officers in nontraditional ways that reflect the needs of the community. Communities, therefore, get to establish their own definition of community policing. This appears to follow the philosophical stand that the local communities must identify the problems and their potential solutions, but the reality is that many communities simply assign the officers to foot patrol or base them in substations without altering the stance toward cooperative problem solving. The placement of officers in the communities is little more than community relations work.

A second problem is that most police training has not changed with the hiring of the new community police officers. The emphasis is still on the criminal code, physical abilities, making arrests, taking reports, and similar factors. Very little training in building coalitions of citizens, problem solving, community advocacy, identifying alternative sources of intervention, working with non-law enforcement agencies, or other key aspects of community policing is provided. Similarly, the criteria for selecting police officers has not changed with the advent of community policing. Working with diverse constituencies to solve problems and building effective coalitions calls for skills that have not been the cornerstone of most selection processes. Different background characteristics and training are needed in community police officers. Without these changes, the simple hiring of staff and calling them community police officers is not enough. Unfortunately, there is no requirement that officers hired under the COPS program receive any specific training or assistance.

This does not mean that community policing does not exist in the United States. Indeed, many communities have embraced the idea and are taking steps to implement community policing in accordance with the philosophy set forth above. The majority of agencies, however, have used the COPS funds as a means of hiring more officers, rather than as seed monies for shifting toward the community policing model. Most of the discussion to follow examines programs and initiatives that attempt to implement true community policing programs.

HISTORICAL BACKGROUND OF COMMUNITY POLICING

Failure of the police and traditional police practices over the last thirty years to curb crime has led to a wide array of strategies aimed at improving the effectiveness of the police. Many strategies have involved moving the police back into the community and enhancing the interaction between officers and citizens. Included in these efforts have been foot patrols, neighborhood or storefront offices, problem-oriented policing, and neighborhood watch programs. An underlying assumption in these efforts was the idea that the public must be more proactive in contacting the police and offering information about crime. The police could then intervene.

One key strategy was the revival of foot patrol. Once a cornerstone of policing, foot patrol was abandoned in favor of automobile patrol due to the perceived advantages of covering a wider area and responding more quickly to calls for help. Evaluations of existing foot patrol, however, reveal a mixed impact on crime (Bowers and Hirsch, 1987; Esbensen, 1987; Police Foundation, 1981; Trojanowicz, 1983). The failure of foot patrol to significantly or consistently reduce crime is somewhat offset by findings that foot patrol does reduce the level of citizens' fear and improve attitudes toward the police (Brown and Wycoff, 1987; Police Foundation, 1981; Trojanowicz, 1983). Variations on foot patrol, such as bicycle patrol and storefront offices, serve much the same purpose. Many of these ideas are incorporated into community policing programs.

Neighborhood or community watch and citizen crime prevention programs are also initiatives from which community policing has evolved. These programs seek to involve the public in efforts to make it harder to commit crime and improve the chances that offenders will be seen and the police will be called. Although the police are instrumental in establishing these programs, their emphasis in the past has been on turning the citizenry into an extended informant network that will call on the police to undertake traditional actions of arrest and investigation. The police recognize that they need the help of the public; they prefer, however, that such help come mainly in the form of surveillance activity. There is little emphasis on identifying the root causes of the problems or the proper solutions.

Police crackdowns, intensive patrol operations, and problem-oriented policing are additional examples of precursors to community policing. Even though police crackdowns and variations on patrol operations appear to be nothing more than traditional policing, they stem from the premise that a specific problem exists and needs to be handled differently from normal practice. For example, a massive police crackdown in one area of New

York effectively reduced drug crimes, robbery, and homicide (Sherman, 1990). Similarly, a police crackdown on a drug market in Lynn, Massachusetts, successfully reduced drug-related crimes in the area (Kleiman and Smith, 1990). Under problem-oriented policing, the emphasis is on identifying a specific problem and initiating an intervention tailored to the unique needs or situation. What differentiates these efforts from true community policing is the fact that police sweeps, intensive patrol, and other actions do not require citizen participation. The police remain the driving force (and possibly the only force) involved in many problem-oriented approaches. The police simply use their traditional powers of arrest to impact the crime problem.

The closest approach to community policing is that of situational crime prevention. As with community policing, situational prevention seeks to identify and attack the basic causes of problems in the community. This endeavor requires the input of various individuals and groups, particularly in the efforts to intervene. The police may be a major player in situational prevention, or they may not be involved in any way. The role of the police will vary by the problem and the solution. This differs from community policing in that community policing assumes that the police must take the lead in the identification of problems, coordination of identifying potential solutions, and overseeing of results.

Clearly, community policing is not a new or unique idea in the United States. The move toward increasing the interaction between the police and the public in nonarrest, noncrime situations appears in a number of initiatives over the past thirty years. What is new is the idea that the appropriate solution to a problem may involve actors beyond the criminal justice system. No single intervention will work for all problems, and no single agency or individual should be expected to solve the problems that cause or allow crime to occur.

COMMUNITY POLICING PROGRAMS

Community policing programs are growing throughout the United States. Wycoff (1995) reports that over 800 law enforcement agencies are involved in some form of community policing. The programs also appear in communities of virtually any size, from small, rural communities to some of the largest U.S. cities (Weisheit et al., 1994; Wycoff, 1995). The following discussion provides four examples of community policing initiatives prompted by both local and federal agencies. It is important to remember that the ambiguity in the definition of "community policing" permits a great deal of experimentation and innovation in problem solving.

Local Examples

Theft from Automobiles

One good example of community policing, which is also used as an example of situational crime prevention, entailed efforts to thwart theft from automobiles at a naval shipyard in Newport News, Virginia. Eck and Spelman (1987) noted that the response to the problem went through an iterative process in which different aspects of the problem were identified and addressed over time. Included in the analyses were the use of maps to identify patterns in offending, the identification of known and potential offenders in the area, and the inclusion of other agencies in the development of long-term responses to theft from automobiles. The outcome of the interventions was a drop in reported thefts of over 50 percent (Eck and Spelman, 1987).

A central component of this initiative was the development and testing of the SARA process for problem solving. SARA involves *s*canning, *a*nalysis, *r*esponse, and *a*ssessment. *Scanning* entails the identification of the problem, issues, and concerns in the community by the police, residents, businesses, or other agencies. After the problem is identified, an *analysis* of the problem is undertaken, coupled with identification of appropriate individuals or groups that should be involved in any response. The *response* differs according to the problem. In some cases the police may have little involvement in the intervention since the identified response may require expertise and abilities that the police do not have. Finally, an *assessment* of the response is needed to provide feedback to any continuing intervention. The entire SARA process focuses on the causes of the problems, not the symptoms. This process is a cornerstone of many community policing and problem-solving initiatives.

Civil Abatement

A second example of local-level community policing involves the use of *civil abatement* procedures to attack problems such as drug dealing. Civil abatement typically involves landlords, health departments, zoning boards, and city or county attorneys in the application of civil (and sometimes criminal) codes. For example, residents, community groups, the police, and other agencies can attack local drug problems by seeking fines against property owners who ignore drug problems in their buildings, by calling for buildings to be confiscated or boarded up when health code violations are found, by evicting tenants who violate the law, or by seeking the demolition of structures that are abandoned and beyond repair.

One well-known abatement project is the SMART program (Specialized Multi-Agency Response Team) in Oakland, California. Initiated in

1988, the SMART program relies on the cooperation of the police, citizens, and other groups in identifying and dealing with drug locations. Although civil court remedies are the cornerstone of the project, the police are a critical part of the process, providing both enforcement and patrol in affected areas of the community. Green (1995) notes that the project has reduced drug activity in the target sites. In addition, the physical condition of the sites, both internally and externally, has greatly improved.

The results of the Oakland program, as well as others, show that civil abatement projects can be successful in building coalitions of citizens and that they are effective in curbing target problems. Residents, however, must be made aware that this process can take a long time, particularly if property owners opt to fight the procedures through the courts. Hope (1994) notes that use of civil litigation may be most effective as a persuasive tool to bring about quicker responses, such as the sale of properties to more responsible owners.

Federal Initiatives

Weed and Seed

One of the most well-known federal initiatives is the Weed and Seed project. This project is aimed at revitalizing communities through a process of "weeding" out existing problems and "seeding" communities with programs and interventions that inhibit future recurrence of the problems. The criminal justice system, particularly the police department, is intimately involved in the weeding portion of the program. The police often undertake concerted patrol operations in problem areas, make arrests, and assist in the prosecution of problem individuals (Conley and McGillis, 1996). Common targets are drug dealers, gangs, and "adult entertainment" businesses.

Once the major problems are eliminated (or greatly reduced), seeding is supposed to take place. Seeding can take a variety of forms, depending on the unique needs of the community. Efforts may involve physical changes in the area, such as the demolition of abandoned buildings, cleaning up of vacant properties, and the erection of buildings for new businesses. At the same time, seeding may require the initiation of educational services for area youths, the institution of employment services and opportunities for residents, and enticing new businesses into the area. Yet another realm for seeding may be to help establish community organizations and increase social interaction between residents. Clearly, the seed portion of the program requires the involvement of diverse agencies and groups beyond law enforcement.

Early evaluation of weed-and-seed initiatives do not show much promise. Typically, only half the effort is successful. Tien and Rich (1994) report that the police are often successful at weeding out the existing problems. The seeding efforts, however, often fail because the participants cannot agree on the appropriate direction the seeding efforts should take. Added to this disagreement is the fact that adequate funds for the seeding initiatives are not available (Tien and Rich, 1994). Kennedy (1996) notes that most efforts are focused on cleaning up neighborhoods (weeding), rather than on improving education, employment, health, and economics in the communities (seeding). What may help is the development of specific seeding plans prior to any weeding activity.

Innovative Neighborhood-Oriented Policing

In 1990, the Bureau of Justice Assistance initiated the *Innovative Neighborhood-Oriented Policing* (INOP) program in an attempt to reduce drug problems in communities. On paper, INOP is a good example of how community policing is supposed to operate. The program explicitly recognizes the need for cooperation between the police and numerous community groups, agencies, and individuals. In addition, it does not propose any single "best" approach for solving drug problems. Each community and program is expected to tailor its own response to the unique situation it faces. Possible responses can include traditional police enforcement tactics, community organizing, education programs, and drug treatment.

Sadd and Grinc (1994, 1996) conducted an extensive evaluation of INOP undertaken in eight sites across the United States. Unfortunately, program participants reported generally negative results for the program. Many efforts appeared to simply displace the drug problem, rather than bring about absolute reductions in drug usage. The authors also noted that the police generally resisted the program owing to poor planning, a lack of clearly defined goals, poor coordination of efforts among participants, and unclear expectations of police involvement (Sadd and Grinc, 1994, 1996). Despite these problems, there was evidence of increased community involvement and cooperation in addressing local drug issues. One of the most valuable outcomes of INOP was the identification of issues and concerns that are often poorly addressed in program planning and evaluations of community policing.

PROBLEMS AND CONCERNS

Community policing has not been universally adopted, largely because of a number of unanswered questions and the underlying philosophical changes being asked of the police. Arguments against community policing often involve issues related to traditional police activity. One concern is the perceived need for quick response time and the expectation that community policing is incapable of such activity (Skolnick and Bayley, 1988; Wycoff, 1995). Another criticism is that community policing is not "real" policing and is a threat to police professionalism (Skolnick and Bayley, 1988).

Much of the problem is due to the lack of familiarity with community policing. Most officers who become involved in community policing claim that the potential benefits far outweigh the drawbacks (Skogan, 1995). What is needed more than anything is recruit selection and training geared toward community policing. Community advocacy and development requires a different type of training and skills from that needed under traditional policing. Officers need to know how to conduct community outreach, build coalitions, and identify problems, not just how to respond to calls for service. At the same time, the public also needs to be educated about community policing and what they can expect of the police under such an approach. Rather than expect high arrest and clearance rates, the community should look for more interaction with the police and others, an improved quality of life, reduced fear of crime, and other positive outcomes.

SUMMARY

A major argument underlying community policing entails the position of police officers in society and the types of actions they currently provide. Evidence shows that the average police officer spends less than 20 percent of his time actually enforcing criminal codes (Lab, 1984). Community policing suggests that police officers can become leaders in the community. Officers should help identify problems, involve residents and others in deciding on potential solutions and implementing those actions, and assist in evaluating the outcome of the efforts. Rather than try to solve societal problems on their own, the police should work with everyone in society to make communities a safer place in which to reside.

CHAPTER 2

CRIME PREVENTION: A COMMUNITY POLICING APPROACH

Maximilian Edelbacher

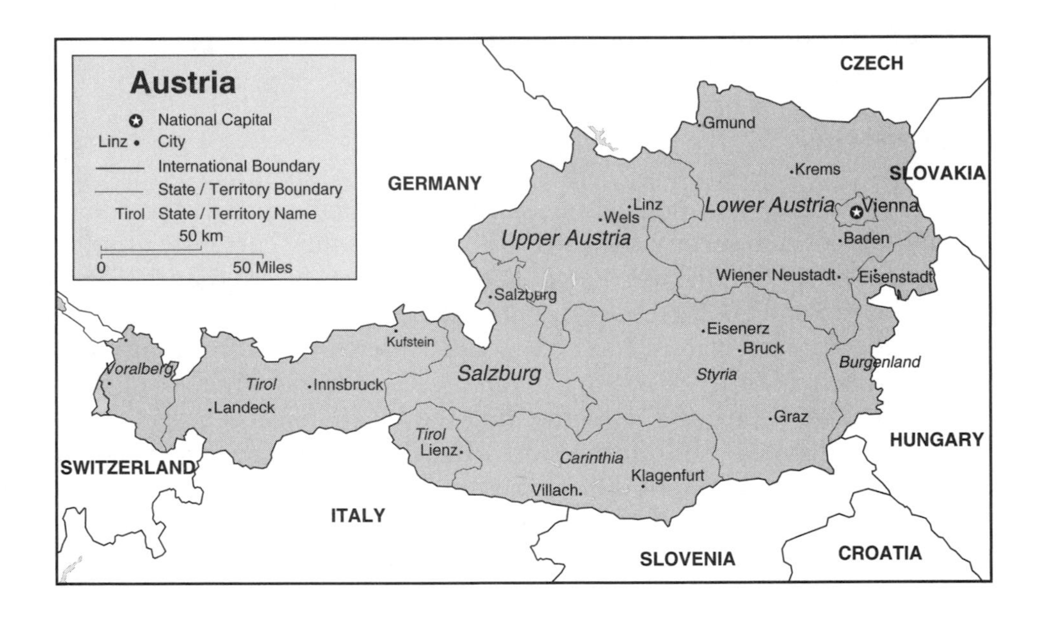

Situated in the very heart of Europe, Austria is a relatively small country with an area of only 83,857 square kilometers (32,378 square miles). Austria is home to eight million people, nearly one million of whom are foreigners who cannot speak the German language or are only partly familiar with it. The capital of Austria is Vienna with about two million inhabitants. It is situated in the eastern part of the country, neighboring Prague, Bratislava, and Budapest. It is widely accepted that more than 100,000 people live in Vienna illegally.

The Austrian Penal Code of 1974 differentiates between serious offenses (felonies) and less serious offenses (misdemeanors). The total number of all criminal acts in Austria has almost doubled since 1975 (see Table 2-1). Figures specifically for Vienna show similar increases.

The Austrian police are organized nationally. The Federal Ministry of the Interior is responsible for the internal security in the country, and the Federal Minister of the Interior is the executive responsible for maintaining public order in Austria. Austria employs about 32,000 police officers. In rural areas, where approximately 5.2 million people live, the federal *Gendarmerie* (federal rural police) are active with about 12,000 officers. In the fourteen largest cities of Austria, with a total of 2.8 million inhabitants, the federal police operate with about 15,000 officers (Institute for Demographics, 1996).

CRIME PREVENTION OR COMMUNITY POLICING

Policing has a long history in Austria. In the past, the duty of police was primarily to guard the emperors and to execute their orders. In fact, the main function of the police had always been to guard the ruling elite.

TABLE 2-1 CRIMINAL OFFENSES IN AUSTRIA AND VIENNA

Austria	1975	1990	1994	1997
Felonies	70,612	101,635	107,868	102,182
Misdemeanors	210,121	355,988	396,700	379,367
Total	280,733	457,623	504,568	481,549
Vienna				
Felonies	31,057	49,620	48,318	43,273
Misdemeanors	63,918	114,114	123,364	115,674
Total	94,975	163,734	171,682	158,947

This picture has changed dramatically in the current democratic society. Increasingly, the role of the police has changed from an instrument of the ruling class to that of servant of the general public. This chapter focuses on this evolution. Crime prevention is an essential part of this chain of development.

Over the past few decades, the job profile of law enforcement officers in Austria has changed dramatically. On the one hand, the job of a police officer is increasingly viewed as a service: The officer is perceived as being there to help the people. On the other hand, an officer, as a highly professional authority, is challenged with combatting organized crime, economic crime, and professional criminals. Police officers try to fulfill both roles. Basic and advanced training and additional courses help the police comply with this job profile.

When Austrians are asked what disturbs them most, the offenses most frequently cited are violence, burglaries, and drug crime. In everyday life, Austrians are confronted with the uncommon customs and ways of life of the many different foreigners who have come to live in the country. Foreigners often dress differently—even seem to smell different—and have different habits that are often considered a nuisance. Police officers must sort out which actions are truly criminal and which are merely annoying. In such cases, the police officer often acts as a sort of Justice of the Peace. The police are constantly challenged to deal with both the "small and big problems" of citizens. Striking the right balance between the two is possible only through close cooperation with the public.

HISTORY, PHILOSOPHY, AND CONCEPT OF CRIME PREVENTION IN AUSTRIA

In the mid-1970s, the police were looking for new approaches to their work. The top police officers in Vienna assumed a key role in importing new ideas. On May 8, 1974, the Federal Ministry of the Interior issued a directive to all Security Headquarters, Federal Police Authorities, and provincial Rural Police Commands to establish the Criminal Advisory Service. At the Federal Police Headquarters in Vienna, the Criminal Advisory Service, a section of the Major Crime Bureau, was placed under the command of the head of the burglary section. In doing its job, the Major Crime Bureau informed people about what should be done to prevent criminal offenses. In this first phase, the leading officials of the federal Police Headquarters in Vienna gave lectures to interested audiences to explain how offenders act and how people can protect themselves against violent crime in their apartments and homes. Older people showed a special interest in this information. Soon, special events for senior citizens were held to pass on this knowledge to an even wider audience. The annual "senior citizen exhibitions" in the Vienna Stadthalle met with approval. Now, 25 years later, they are still well-attended.

The British Model of Crime Prevention

A visit to the British police by Austrian authorities in 1977 added strength to this plan. The British crime prevention model, "to serve the community," became widely accepted in Austria. Already in the 1970s, and to an even greater extent in the 1980s, criminological research was able to prove that both special and general preventive measures have their limits. Neither with treatment and therapy nor with deterrence and repression can the problem of crime be solved. Obsolete police measures had to be given up (for example, detection can be increased only to a minimal extent and will not solve the problems) (Dölling and Feltes, 1993). In the 1970s, community policing projects were still uncommon in Austria. Because the police in Austria are organized centrally, the development of community policing was inconceivable. All sides of police management tried to increase the confidence of the population in the work of the police.

Development of Crime Prevention: The Security Guard

Several changes have led to increased crime prevention. In 1977, "contacting officers" were introduced at the Federal Police Headquarters in Vienna. The job of the contacting officers (police officers in uniform) was to make direct contact with the people in their district. Business people and the elderly were the main contact persons sought out by these officers. Since 1983, the Security Guard has intensified its patrolling activities. This measure was supposed to reduce the fear of crime among the population. In 1984, "youth contacting officers" were introduced at the Federal Police Headquarters in Vienna. The job of these specially trained officers was to contact young people and help them cope with their problems. In the course of their activities they deal with road safety education, training for drivers of motor scooters and motorbikes, and caring for youth centers and clubs.

All these measures were first introduced in Vienna and have been duplicated in other Austrian cities. It cannot yet be determined to what extent these measures have contributed to overall safety in Austria. Between 1970 and 1995, the crime rate doubled in Austria. Public confidence in the police, however, changed little between 1973 and 1989 (39 percent of those questioned thought that the police were taking sufficient care of the safety of the population in 1973, compared with about 41 percent in 1989) (Austrian Federal Bank, 1998).

COMMUNITY CRIME PREVENTION: THE NEW ROLE OF THE POLICE

On May 1, 1993, the Security Police Act went into effect. The act serves as the legal basis of all powers and duties of the police toward the citizens. This act creates a new relationship between the police and the citizens, balancing power and duties between citizens and the police. It also forms the basis for crime prevention. Crime prevention is primarily carried out by the Criminal Advisory Service, with criminal counseling being a part of crime prevention.

The goal of crime prevention is to deter offenses punishable by the court. There is a strong interaction between repression and prevention. The police consider themselves authorized to carry out preventive measures, not because of their legal mandate, but because of their special experience and knowledge.

The daily confrontation with all forms of crime inevitably leads to considerations of prevention. The police are accepted among the population as a competent partner in security matters. Crime prevention is a task incumbent on the whole society. Only local authorities and other institutions, as well as parents and teachers, can have the influence needed to create conditions that will reduce crime.

Prevention is the task of all police employees. For instance, the police officer on patrol may tell a car driver that his car is not locked, an investigating criminal officer may give advice to the neighbors of burglary victims on possible technical safety measures, and employees at the passport office may inform people of safety measures during their holidays when they come to get their passports.

Apart from these activities, the demand for experts who deal exclusively with crime prevention has developed. Extensive experience, competence, and technical knowledge can be obtained only by intensively dealing with the topic of prevention. To keep up with recent developments in prevention, it is necessary to gather comprehensive information, which takes time and requires specialization. Practical experience shows that, in some situations, it is essential that contacting officers should not come from the investigation department.

Organization of the Criminal Advisory Service

The Criminal Advisory Service of the Vienna Police is the central authority for all prevention work (Jedelsky, 1998). In the past, the problem of crime prevention and community policing in Austria was based on the hierarchic structure of the Federal Austrian Police and *Gendarmerie*. A new philosophy had to be created. The new idea was that "the police should come to the people, not that the people should have to go to the police." There are four sections of the Criminal Advisory Service:

- Group for prevention of property crime
- Group for prevention of violence against children, youths and family
- Group for prevention of sexual offenses
- Group for prevention of addiction

The central department consists of male and female police officers, lawyers, criminal investigation officers, security guard officers, and other employees. These people comprise the "prevention team of the Viennese police force." Additional officers perform preventive tasks in other central departments and at the district police stations. The Office of the Criminal Advisory Service regularly organizes training courses and information lectures for officers. All officers of the federal Police Headquarters in Vienna are trained in matters of the Criminal Advisory Service as part of their basic training.

Measures of the Crime Prevention Program

Crime prevention is the main focus of all government and private efforts aimed at a reduction of criminal offenses. Primary, secondary, and tertiary prevention can be distinguished. *Primary prevention* combats crime at its root. Points of approach are social politics, strengthening of legal awareness, the explanation of norms, and the creation of incentives for compliance with the law. *Secondary prevention* combats crime on the surface using classic measures like police presence, changes in the opportunity structure for an offense, self-protection, improvement of technical preservation methods, and increasing the risk for offenders. *Tertiary prevention* includes offender-oriented activities, such as offender-victim leveling, social training courses, and resocializing measures.

The Criminal Advisory Service focuses on information in the secondary area, but it is also the initiator and driving force in other areas of prevention. The service encompasses a wide range of efforts, including lectures, individual counseling, observation and analysis of crime scenes, attempts to solve individual problems, personal protection, object protection, and advice regarding safe living habits. They research and analyze crime, sensitize target groups, cooperate with other institutions challenged with prevention (such as schools and municipalities), build platforms for information exchange, participate in commissions and working groups, offer self-defense courses, test security technology, deal with problem groups (like youth gangs and soccer hooligans) and provide antiviolence training. Clearly the activity is quite broad.

Information on crime prevention is not reserved only for individuals. Companies, authorities, and other institutions can also turn to the Criminal Advisory Service. Consultations often are held on site. State-of-the-art security technology, which is partly subjected to testing, is on display. Officers of the Criminal Advisory Service are represented on various commissions and standards committees. The security industry, insurance agencies, and banks have come to accept the Criminal Advisory Service as a competent partner. Often, activities in the high-security arena take place, and comprehensive security schemes are required. The Criminal Advisory Service offers its services according to the motto: "Of course, you can hope never to become the victim of a crime, but you can do something to avoid it." The services are offered to prevent theft of property; violence against children, teenagers, and family; sexual offenses; and addiction.

Prevention of Violence against Children, Young People, and Family

One group in the Criminal Advisory Service deals with the prevention of violence against children and teenagers and the prevention of domestic violence. The first and foremost task is to detect problems in certain neighborhoods, parks, recreation grounds, and other meeting places. Knowledge of the local scene is essential. The second task is to inform other institutions so that they can become active. Apart from these efforts, there are attempts at sensitization through participation in problem-related events, providing information and creating an awareness of problems. This prevention group organizes soccer and street-ball games and antiviolence training courses, sets up contact between hostile groups, takes care of fan groups or individual persons, and presents meaningful alternatives. By explaining the legal situation (for example, often young people are not aware of the significance of a "robbery"), they attempt to achieve an awareness of problems.

It is also the task of this group to address and discuss the problem of violence against women and children. In this area, cooperation with other institutions is particularly important, and the Criminal Advisory Service considers itself a coordinator and contact body for giving advice to individual people and organizing social or medical help.

Prevention of Sexual Offenses

Violence against women, in particular offenses against sexual self-determination, and sexual abuse of children constitute another area of preventive work. It has been proved that the number of unreported cases is particularly high, since about two-thirds of the offenders come from within the victim's family or are known by the victim.

The advice given by the police primarily refers to unknown offenders, where appropriate preventive behavior and defense actions are promising. By issuing their recommendations, the police do not intend to restrict the personal freedom of women. However, certain precautionary measures are part of safe and conscious behavior, and this behavior also sets an example for children or younger siblings (for example, a definite *no* to hitchhiking). When trying to prevent criminal offenses, security must take priority over personal freedom.

With offenders who are family members or friends, it is difficult for the police to address the target persons. Here prevention can only stem from a change in awareness. The principle, "Anyone who looks away is also guilty," has to be stressed in order to change the attitude of witnesses. Offenses have to be examined thoroughly in order to develop strategies for prevention. The Criminal Advisory Service is used for this purpose.

Women who take the responsibility for their own safety react better and more self-confidently in dangerous situations. The main thrust of preventive work, therefore, is to strengthen this self-confidence and teach certain defensive techniques. In self-defense courses the women are confronted with problem situations, certain defense techniques are practiced, and self-defense weapons are tested. Children are sensitized to sexual abuse by means of similar strategies. An awareness of the problem is created so that children can protect themselves.

Prevention of Addiction

Addictive behavior in society is the greatest challenge for the Criminal Advisory Service. The use of legal and illegal drugs has increased alarmingly over the past few decades. Prevention aims at strategies to prevent addiction to illegal drugs, as well as legal drugs like alcohol, tobacco, and the abuse of medical drugs. Addictive behavior has also increased. The onset is often characterized by abuse of alcohol and medical drugs.

By providing information to parents, teachers, and young people, everyone will be made more aware of the problem. Young people should be given the opportunity to pursue their own goals and should be taught to overcome personal conflict constructively. Experience shows that prevention does not function by "trying to talk somebody out of something" but only by offering new and better alternatives. It makes no sense to tell young people that drugs are dangerous—they know this. We must offer new perspectives.

Mere deterrence has failed, as has detailed information as to the negative effects of drugs. Because young people tend to take risks more readily, stressing these risks may produce exactly the opposite effect because of innate curiosity, experimental behavior, and a thirst for adventure. Experience in Germany shows that professional preventive work in the field of narcotic drugs has been accepted by young people. The Austrian Criminal Advisory Service holds individual consultations and organizes exhibitions, workshops, antidrug discos and lectures for pupils, parents, and teachers.

The Chances for Success

Are crime prevention methods promising? Can any success actually be measured? Basically, it is very hard to measure how often a criminal may avoid a car featuring a clearly visible antitheft device, or whether a house was not broken into because there was a clearly visible alarm system. Studies, however, show that prevention makes sense. According to a study at the University of Graz, 70 percent of burglars were deterred by alarm systems, and as many as 50 percent were deterred by dogs (Krainz, 1994). The German crime prevention program has been in force for twenty years, and it has been established that attempted burglaries increased from 19 to 30 percent over a ten-year period, since the offenders tend to fail in their attempts more and more often. Sixty percent of the attempts failed due to technical security devices, and the remaining 40 percent failed because they were interrupted, possibly by watchful neighbors.

The Media and the Criminal Advisory Service

The Austrian media maintains a supportive, positive attitude toward prevention. Many television programs publicize information on crime prevention without charge. Organizers provide free floor space at various exhibitions. A journalist's prize for outstanding contributions to crime prevention has been awarded by the Federal Police Headquarters in Vienna. This award is financed by outside sponsors. The high degree of familiarity with the Criminal Advisory Service has to be attributed mainly to articles in the newspapers. Exposure on television or in the print media always cause considerably more visitors to come to the Criminal Advisory Center. For instance, three reports on television in November 1994 brought 82 percent more visitors to the Criminal Advisory Center.

Community Policing in Vienna and Schwechat: Examples of a Breakthrough Strategy

An analysis of actual criminal activity reveals that Vienna is among the safest cities in the world. Subjectively, however, the situation is not as positive. Many citizens, particularly older people, are frightened. Many people are calling for a stronger police force and the use of private guard services and home guards.

The tension between objective and subjective safety is reflected in the problem of juvenile delinquency. Almost every day there are reports in the media about school violence or drug-related offenses. The personal sense of safety does not correspond with the objective situation of actual crimes committed (which is not as bad).

Based on international experience, more police will not automatically cause a decline in crimes. More police cannot produce more security. The legend of the "district police officer" has become obsolete. The nice policeman with the waxed mustache, the measured step, and the stern look but soft heart has been replaced by the young sportive "security expert" dressed in a police uniform (Mahrer, 1995).

In the town of Schwechat, Lower Austria, the first approaches toward community policing were made in 1991. The Security Panel Schwechat was founded as a local pilot project of the Criminal Advisory Service Schwechat. The Hietzing district of Vienna founded an association known as the Partnership Security in Hietzing in 1991. The Federal Police Headquarters in Vienna also decided to pursue a community policing strategy. At the beginning of 1994, a working group commissioned by the former chief of police dealt with this topic and was instructed to prepare a project proposal. In the same year, Vienna developed the Vienna—Safe City project to ensure that all activities would be coordinated at the district level to avoid duplication. Cooperation between the Federal Police Headquarters in Vienna and the municipal government allows community policing projects to be worked out at several levels. In the course of these projects, "security panels" on the Schwechat model were installed in the individual districts.

Can a Community Policing Strategy Succeed?

The chances for success of community policing lie in the prevention of crime at its roots. As an alternative to the demand for more police, people are directly involved in the process of ensuring on security. The readiness of citizens to accept responsibility and their visualization of the actual trouble imposed by crime will improve the overall sense of safety.

The new form of real cooperation between the community and the police gives rise to a profound relationship of trust. Police work is not only limited to the handling of emergencies ("they come only if something has already happened") but becomes more comprehensive and professional. This comprehensive police work provides an opportunity to redefine the job profile of police officers and thus the chance to gain satisfaction on the job (Mahrer, 1995).

What does community work actually consist of? This question was posed to a working group at the Federal Police Headquarters in Vienna. Five pillars of community police work were identified (Mahrer, 1995).

The Five Pillars of Community Police Work

The first is the reliance on *joint* security. The problems pointed out by the citizens must be solved using the capabilities of the police. Under this scheme, possible solutions as well as the limitations of police work will be made clear by including other institutions. Independent associations and institutions (for example, home guards) can enhance security by increasing the responsibility of the citizens and preventing undesirable developments.

A second pillar is developing partnerships without anonymity. Voluntary officers who are put in charge of the project and are in direct contact with the community do not act anonymously. Instead they use their own names. Such openness creates trust and promotes cooperation between the various partners. This form of making contact by the police is supported by self-confidence, adaptation of the training system, and encouragement by top officers.

Linking to other institutions is an important third improvement. Through regular, personal or event-related contact and through the exchange of specific information between community initiatives and the police, concerns can be solved jointly with a common objective in mind. Thanks to enhanced understanding of technical matters, the police are in a position to increase their professionalism and to concentrate on police-relevant subjects in their work (for example, the link currently planned by the Federal Ministry of the Interior for the formation of "intervention units" to handle domestic violence).

The fourth pillar is decentralization and internal dialog. Because of the great variety of local problems, security work can be planned and implemented by means of delegating authority. The principle of delegating responsibility has to be observed in both local areas and technical activities. Delegation of responsibility means an accepted loss of power for some, but motivation for many. This requires a high degree of discipline, a constant exchange of information, and the acceptance of centrally defined conditions. Personal contact with the individual citizen is the prerequisite for successful cooperation based on partnership.

Finally, clear structure in small units is a key factor. Through decentralized departments, police work becomes more effective on the local level. Effectiveness is guaranteed by regularly changing patrolling areas and by the stimulation of the community itself. This patrol scheduling is based on steady contact with the population, as well as on a mutual flow of information between local and central bodies. On the whole, this guarantees that the police can keep up to date with "peripheral dark fields," that they can become active during egregious situations, and that they do not have to wait to act only during an emergency.

Lessons of the Hietzing Experience

In the course of putting theory into practice, many problems arise. At the annual meeting of the Criminal Advisory Service in September 1994, it was revealed that problems in Hietzing applied to all of Austria, and would serve as a benchmark for learning about community policing. Hietzing is a largely upper-class district of Vienna. In August 1993, the Community Policing Project got under way by providing information and training to project managers (police station commanders and contacting officers) in cooperation with the district manager. This consolidation phase between community representatives and police officers was vital in building mutual acceptance. Despite this success, many police officers still distrust the new approaches. However, those actively involved in the introduction of the project are able to tackle the matter with less prejudice.

Since the autumn of 1993, regular meetings have been held in shops designated to function as key locations to handle security. A page in the monthly district journal is devoted to security. In-school project weeks have been carried out in cooperation with the security guard and criminal officers. Regular contact with youth centers, as well as direct contact with citizens in focal spots of the district, have brought about an end to open drug activities in endangered areas (for example, Kennedybrücke). In the autumn of 1995, a pilot trial of "mobile police" was introduced (also conducted in five other districts of Vienna). *Mobile police* refers to a moving police station. It consists of an adapted police car manned by three to five officers and assigned to critical areas. These officers provide the same services that are usually performed at the police station. The use of the mobile police is supposed to emphasize service orientation and good contact with the community.

Risks and Consequences of Community Policing Strategies

One risk is political monopolization. At times it has not been possible to reach a consensus among all the political parties involved. The consequence is that the police have been forced to take a neutral stand. They are the contact persons for all citizens and, accordingly, for all democratic groups acting within the scope of the law.

The tendency toward centralization is another risk. Instead of delegating responsibility, "tasks" are delegated. "Networking" remains a catchword. The need is for actual and complete decentralization. Only then will it be possible for "community" security work to perform locally. The central departments have to take on the important function of control and guarantee cooperation with community institutions in order to integrate networking as an essential factor of community security work. After initial help from centralized bodies, local networking must be put into practice on a daily basis. Networking ordered from "above" is destined to fail.

Successful management becomes difficult because of a lack of specific success criteria. As a consequence, systematizing and planning of projects must proceed with objective planning and the use of information.

Another risk is a lack of motivation due to insufficient training and consolidation. Before new strategies are introduced, it is indispensable to "work off" prejudices and justified points of criticism. The police employee or high-ranking official who reads in the paper that, as of now, community policing projects are being instituted "is no longer able to come along." Acceptance will take place only through training and reeducation.

Finally, the police must overcome the attitude that they should command while the people merely stand by and watch. Consequently, the police must stimulate activities that allow security to become a concern of the community. The police can remain merely problem solvers for exclusively police-related problems, or they can act as links to those actually responsible for the problems.

Crime Prevention and Community Policing in Austria

The Community Policing Project has been imitated in Vienna by several districts. In the other thirteen major cities of Austria, there are similar projects. Crime prevention is a major goal of all cities. But community policing projects are handled differently in Vienna and the other cities from the way they are handled outside the cities. In the rural areas, a very extensive communication network has always existed between the *Gendarmerie* and the people. People know the Gendarmes, and the Gendarmes know the people. Mainly they trust each other, and there is much more communication and mutual cooperation about problems than in the cities.

FUTURE OF CRIME PREVENTION IN AUSTRIA

The acceptance of crime prevention methods in Austria, especially in Vienna, is widespread. In fact, crime prevention has a relatively long tradition in Austria. Because crime has doubled in the last twenty years, people feel insecure, they long for security and, as a result, support crime prevention.

Many complications are involved in introducing community policing methods. The new system is integrally different from the old. In some parts of Austria and Vienna, people accept these new approaches. In other areas, they are not interested in a modernization of the relationship between the police and citizens. There is a big gap between the people's subjective sense of insecurity and the objective reality of crime in parts of Vienna. On the one hand, Austria is relatively safe compared to other countries in Europe and in the world as a whole. On the other hand, many people feel insecure because of the open borders, the migration movements, and the new political and economic situation created by the European Community. Crime prevention and community policing are viable approaches to the future of effective policing in Austria.

CHAPTER 3

COMMUNITY POLICING IN THE ROYAL CANADIAN MOUNTED POLICE: FROM PREVENTIVE POLICING TO A NATIONAL STRATEGY FOR CRIME PREVENTION

R. J. Laing
L. T. Hickman

The Royal Canadian Mounted Police (RCMP) is Canada's national police force. With 15,142 sworn officers and an additional 5,319 civilian members and public servants, the RCMP is the largest police organization in Canada (Statistics Canada, 1996). The RCMP is organized pursuant to the authority of the R.C.M.P. Act. In accordance with the act, the RCMP is headed by a Commissioner, who, under the direction of the Solicitor General of Canada, is responsible for the control and management of the Force. The RCMP consists of 14 divisions, with headquarters in Ottawa, Ontario. Each division is managed by a commanding officer and is alphabetically designated. Divisions roughly approximate provincial boundaries. The divisions have 52 subdivisions and more than 700 detachments, ranging in strength from 1 to 380 officers (RCMP, 1996).

The mandate of the RCMP is "to enforce laws, prevent crime, and maintain peace, order and security" (RCMP, 1994). The police service delivery of the RCMP is based on the community policing philosophy. This comprehensive organizational and operational approach provides for the reduction of crime and attempts to meet the social needs of the communities it serves.

Through Shared Leadership Vision,[1] restructuring, and the continuous move toward full implementation of the community policing philosophy, the RCMP is better able to anticipate and respond to the incredible change that has impacted every agency within the police universe. Widespread change within the RCMP has made it possible to align its service delivery with priorities at all levels of government and communities.

Change has created additional opportunities to develop strategic partnerships to accomplish problem solving in many areas, including how to prevent crime. When fully implemented, the underlying philosophy of community policing and the use of problem-solving tools have eclipsed and replaced the program-oriented crime prevention/police community relations model that the RCMP relied on for several decades.

This chapter discusses the RCMP community policing philosophy as a service-delivery model that provides a community platform to solve problems and prevent crime. Beginning with a brief profile of Canada and a review of the literature, the chapter attempts to capture the essence of the history of the RCMP, with a focus on the delivery of crime prevention/police community relations services and the recent development and implementation of the RCMP community policing philosophy. Difficulties in implementing community policing are also discussed. This chapter defines the RCMP community policing philosophy, highlights the alignment of this police service delivery with the priorities of government and the National Crime Prevention Council, and explains the RCMP partnership role in advancing the national crime prevention strategy. The chapter concludes with a prediction of what the future may hold for crime prevention through community policing within the RCMP.

REVIEW OF THE LITERATURE

The preventive-policing, service-delivery orientation of the RCMP during its formative years radically changed to a professional model during the technological era. This had the effect of isolating the RCMP from communities and its partners in public service (Horrall, 1974; Jennings, 1974; Lotz, 1984; Trojanowicz, 1994). The program-oriented crime prevention model developed by the RCMP may still serve a purpose in close-knit smaller communities. However, there is a growing need and demand to offer comprehensive, long-term, integrated crime prevention service with a priority focus on children (Ward, 1997; Horner, 1993; National Crime Prevention Council, 1997; Trojanowicz, 1994).

Contrary to conventional wisdom, program-oriented crime prevention initiatives are relevant in close-knit, small, intimate communities with low crime rates and high levels of informal and symbolic social control mechanisms (Ward, 1997; Collins, 1982). In 1988, the RCMP returned to the community policing philosophy (Leighton, 1994; RCMP Corporate, 1998; RCMP Pony Express, 1995). Impediments to community policing include implementation difficulties in philosophy, structure, strategy, and culture. These issues became axiomatic in the police community, including the RCMP (Greene et al., 1994; Roberg, 1994; RCMP, 1998). Overcoming community policing impediments is a work in progress. The measures taken to overcome these issues include community policing training for employees and community leaders, enhanced communication, futuristic management concepts, and shared leadership vision (RCMP, 1994; RCMP Corporate Management, 1995; RCMP, 1998).

A key component of community policing is to work closely with the community and other agencies to address root causes of crime and social disorder (RCMP Corporate Management, 1995; Ward, 1997; Leighton, 1994; Brannigan, 1997). This position is shared by the National Crime Prevention Council, which adds that key partners in the community must mobilize and take action to develop skills, tools, and knowledge needed to provide crime prevention services for children, prenatal to age 18 (National Crime Prevention Council, 1997). Community policing has positioned the RCMP and its key community partners to support and become involved in the national crime prevention strategy. In fact, the RCMP is already following the recommendations of the National Crime Prevention Council in several communities (National Crime Prevention Council, 1997; RCMP, 1998). Through alignment, cultural change, and improved communication, the RCMP is moving toward full implementation of the community policing philosophy and will accept a key partnership role in developing and implementing a national crime prevention strategy (Murray, 1998; National Crime Prevention Council, 1997).

COUNTRY PROFILE

"Most of Canada's 10 million square kilometers are uninhabited. Indeed, about three in four Canadians live in a widely-spaced string of cities close to the border with the United States" (Statistics Canada, 1996). The July 1, 1995, estimate of Canada's total population was 29,606,100. "Far from being spread out evenly across this vast space . . . three-fifths (more than 18 million people) live in the 25 largest census-metropolitan regions of the country" (Statistics Canada, 1996:60). The population density ranges from more than 7,753 people per square kilometer in Montreal North to one person per 60 square kilometers in the Northwest Territories. The aging "baby boomers" are having a significant demographic influence in Canada, where the 3.4 million seniors in 1993 are expected to reach 11 million by 2041 (Statistics Canada, 1996).

Politically, the Canadian government is a federation, meaning that the broad powers of government are distributed between a central (federal) government and 10 provincial governments, each exercising its own powers, as outlined in the Constitution. The Northwest Territories and the Yukon are governed under delegated federal legislative jurisdiction.

The justice system and police in Canada attempt to uphold the goal of the Constitution in providing "Peace, Order and Good Government."

> Balance is the hallmark of our system . . . our governments, courts and police—the makers, interpreters and enforcers of our laws—are dedicated to upholding the fundamental values of Canadian society, while at the same time addressing the needs of a constantly changing country (Statistics Canada, 1992:468).

> The Royal Canadian Mounted Police, through the Ministry of the Solicitor General, is the primary federal law enforcement agency responsible for enforcement of most federal laws. All provinces and territories, except Quebec and Ontario have entered into contracts with the RCMP to enforce criminal and provincial laws. The Ontario Provincial Police (OPP) and the Quebec Provincial Police (QPP) provide provincial policing services to their respective provinces, while the Royal Newfoundland Constabulary shares provincial policing duties in Newfoundland and Labrador with the RCMP. Legislation in most of the provinces/territories makes it mandatory for cities and towns to maintain their own police force . . . municipalities may create their own police department or contract police services from the RCMP, OPP or QPP (Statistics Canada, 1992:8).

> Canada has one of the highest crime rates among industrialized countries. Although our rate of violent crime is significantly lower than that of the United States, our level of property crime is actually very close to that of the U.S. In 1993, 24% of Canadians were victims of at least one crime or attempted crime.

> In 1994, 2.6 million Criminal Code incidents were reported: approximately 12% were violent crimes, 58% were property crimes and the remaining 31% included crimes such as arson, prostitution and restricted weapons offenses. . . . In 1994, homicides in Canada occurred at the lowest rate in 25 years; 596 incidents of first and second degree murder, manslaughter and infanticide, eight in ten homicides were solved by the police, 86% of homicide victims were killed by a spouse, other family member or by an acquaintance, and only 13% by a stranger. In 1994, 58% of youths (age 12–17 years) charged with Criminal Code offenses were charged with property crimes, 18% with violent crimes (mostly minor assault), and 24% with other crimes. Youth accounted for 30% of those accused of property crimes, although youth only account for approximately 8% of the population (Statistics Canada, 1996:468–469).

HISTORY OF RCMP CRIME PREVENTION

The Formative Years

The genesis and spiritual beginnings of the North-West Mounted Police lie in the march west and the first ten years of policing the Canadian frontier (Horrall, 1974:13–14). The Canadian government took a direct hand in territorial law that emphasized preventive, rather than punitive, measures (Horrall, 1974). The mission of the RCMP was to eliminate the whiskey trade, prevent or solve problems before they reached a crisis stage, and bring law and order to the frontier in order to enhance settlement. The "main attributes of the Mounted Police law were [its] simplicity and authority" (Jennings, 1974:54). With a clear mandate and no organizational or legal impediments, the mission of the Mounted Police was perfectly aligned with government policy and the interests of the aboriginal and nonaboriginal communities.

The "Mounted Police philosophy of preventive law enforcement" contributed to "gaining the confidence of the Indians" (Jennings, 1974:63) and was well accepted by the growing number of ranchers and others who began to settle in the Canadian frontier. The homogeneity of the police and settlers did not prevent strict law enforcement within the ranching community. In fact, strict prevention was used extensively by the Mounted Police to "suppress the usual causes of frontier violence, (in an effort) to prevent serious crime" (Jennings, 1974:63). The tradition of being members of the communities they policed and the service delivery orientation of preventive law enforcement, delivered with fairness and common sense, was developed early in the history of the Mounted Police.

The Technological Era

"As the RCMP became older and larger, it lost the flexibility that marked its early style and became more rigid and bureaucratized. By the early 1970s, it had 4500 regulations" (Lotz, 1984:9). The period between 1960 and 1985 brought incredible technological and cultural change to the RCMP. The accumulative effect of the technological advances in equipment and management systems actually left the RCMP behind—behind a steering wheel in a police car, behind an impersonal telecommunications network, behind a desk, and behind a cluster of impediments to effective policing, including professionalization, specialization, institutionalization, and bureaucratization. The RCMP became a leader for modernization, but technology had the effect of isolating the police officer from the community and encouraging reactive police methods. The RCMP began to take a very narrow view of its role as a peace-keeping force. The police culture changed. Respect was given the officer who could attend and clear a complaint in the shortest time, collect the most evidence, charge the most offenders, and put the greatest number of people in jail. All the audit and management processes supported this paradigm. If a front-line constable solved a problem in the community and concluded an investigation without laying a charge, audit and management systems would demand supporting rationale and the file was very closely scrutinized.

As the RCMP was drawn to the incident-driven, reactive policing model, it was distanced as a police force from the community. Communities were told not to worry about crime. After all, it was the law enforcement officer's job to investigate crime. Communities were told their help was not needed. All that was needed was more money to hire more law enforcement officers. The funding sources obliged, and more officers were hired and trained to become specialists. Specialization resulted in losing touch with the community. In fact, the community, or clients, became faceless and unknown, contributing to even greater isolation from the community. Simultaneously, the RCMP front-line commitment to community problem solving and preventive policing was diminished and the unofficial team of public service providers became increasingly isolated from the community and one another, "reducing opportunities for them to act as informal agents of social control" (Trojanowicz, 1994:258).

Many members of the RCMP considered crime prevention an option they could practice if they had time. For some members the job of preventing crime had become a specialized function of the crime prevention/police community relations (CP/PCR) officer. Typically the CP/PCR position was not pivotal in the RCMP hierarchy and, with few exceptions, the members selected to serve in these positions were not among the organization's elite. The cooptation of crime prevention and police-community relations duties may have contributed even further to misunderstanding the importance of crime prevention, both within the RCMP and in the communities.

During this era, the RCMP CP/PCR branch in Ottawa became the central repository for more than "200 programs to assist communities in preventing crime or reducing fear" (RCMP, 1996).[2] The effectiveness of this program-oriented crime prevention service-delivery model, is arguable.[3] The body of knowledge in criminology against program-oriented crime prevention recognizes the importance of responding appropriately to the needs of the communities.[4] The RCMP provides police service in the context of small towns in predominantly rural locations. These communities typically have low crime rates and very well established, highly effective, informal (even symbolic[5]) social control mechanisms that prevent criminal and other undesirable behavior. In these communities, program-oriented crime prevention may be appropriate in terms of supporting the more comprehensive underlying social control mechanisms that are present.

Of the RCMP members who "elected" to opt in and were committed to community crime prevention, few had the benefit of crime prevention training and many made their contribution in off-duty hours, on voluntary (unpaid) overtime. Many officers were not given the tools, skills, or knowledge to do the work. The RCMP members also were without strong leadership in the area of crime prevention and without the guidance of a comprehensive national strategy for crime prevention.[6]

A Return to Community Policing

It may be argued that the RCMP formally adopted the community policing philosophy and community problem solving as a result of the overpowering influence of its American neighbor (Leighton, 1994). The RCMP and others, however, view the transition to the community policing philosophy as a "return to their roots . . . after a few decades of flirting with the professional policing model" (Leighton, 1994:21). Regardless of the reason for adopting or returning to the community policing philosophy, the commissioner's Directional Statement in January 1990 made community policing a strategic goal and provided a brief definition and rationale for moving toward implementation of this service delivery model:

> Community policing is not a self-contained program but a method and style of delivering most police services. . . . community policing emphasizes the broader responsibility of the police as part of a community to maintain order and reduce fear in that community. This broad responsibility is fulfilled by demonstrating local accountability and treating the community not as a passive recipient of police services but as an active agent and partner in promoting security. It requires establishing as operational priorities those problems that disturb the community most, adopting a pro-active, problem-solving approach and measuring effectiveness by the degree of public cooperation received and by the absence of crime and disorder in a community (Inkster, 1989).

IMPLEMENTING RCMP COMMUNITY POLICING

With a belief that function follows structure, the Community, Contract and Aboriginal Policing Services Directorate was established in 1991 with a mandate to bring about the RCMP transition to community policing. The sophistication and complexity of the tasks at hand were underestimated and "the period between 1989 and 1992 proved to be essentially unremarkable, as relatively few significant changes actually took place. . . . The reason was quite simple: the plan lacked a methodological process that could transform the (community policing) principles into practice" (RCMP, 1998).

Community policing is not a program, it is an "organizational philosophy and management style that promotes a partnership between the police and the community, sharing in the delivery of policing services" (RCMP Corporate Management, 1995:3). The guiding principles that put this philosophy into practice include a partnership between the police and the community for the purpose of identifying and resolving problems, empowerment and decision making at the service-delivery level, risk taking and risk management, interagency cooperation, reducing the fear of crime, and accountability to the community (RCMP Corporate Management, 1995). When community policing is effectively implemented, members "routinely engage in problem-solving activities and work closely with other agencies to address root causes of crime and social disorder" (RCMP Corporate Management, 1995).

The difficulties experienced by the RCMP in their attempts to implement community policing are not unique. The RCMP simply did not appreciate, predict, or plan for the immense and complex changes that were needed across the organizational structure, the police culture, and service-delivery philosophy.[7] The widespread difficulties experienced by most police agencies in implementing community policing have become axiomatic. Community policing impediments, therefore, are well known in the police community. Some of these impediments are listed below (RCMP Corporate Management, 1995; RCMP, 1998).

Philosophical Impediments

The community policing philosophy was not communicated effectively to the membership and the community. In the beginning, there was no formal training, resulting in an overwhelming lack of understanding, knowledge, and commitment at all levels, including the community. The community policing approach required change from the linear bean-counting, law enforcement specialists to the holistic problem-solving generalist.

Structural Impediments

Community policing is as much a blueprint for organizational and management reform as it is an attempt to reunite the police and the community. The structure of the RCMP must support the front-line delivery of community policing. Structural impediments have included restrictive policies that stifle risk taking, innovation, and creativity and call for excessive levels of supervision, which is an impediment to moving decision making as close as possible to the place where the police service is delivered.

Strategic Impediments

The internal strategies necessary to support the delivery of community policing have not been effectively communicated to middle and senior management. This lack of communication is accompanied by a general lack of direction by middle to senior management. The decision to advertise successful initiatives, rather than impose "one size fits all—from the top down" programs, is supported by the membership, but the sharing of ideas and best practices is not conveniently available to date.

Cultural Impediments

Organizational theorists conclude that it usually takes between ten and fifteen years for fundamental changes to become institutionalized. The cultural belief that the RCMP has always done community policing is an impediment because it encourages the status quo. This belief does not consider the need to share in the delivery of police service.

OVERCOMING IMPEDIMENTS

Overcoming community policing impediments has challenged many police agencies during the past ten years. The RCMP has attempted to overcome the impediments through various measures.

Training Members, Employees, and the Community

In March 1994, the first RCMP community policing training was provided in Lethbridge, Alberta. During the following six months, all employees of Lethbridge subdivision and several community advisory committee volunteers and community leaders also were trained in the community policing philosophy. With some revision, a new RCMP training model (CAPRA) was developed in 1996. Based on Goldstein's *Problem-Oriented Policing*, (1990), the CAPRA problem-solving tool focuses on knowing direct and indirect *Clients* and their expectations, *Acquiring* and Analyzing Information, establishing and maintaining *Partnerships*, developing a service, protection, enforcement or prevention-based *Response*, and following up with an *Assessment* of results. Cadets are provided an extensive training orientation in community policing and the CAPRA problem-solving model. Community policing and CAPRA training also have been provided across Canada to community members, employees, and community leaders.

Communication

A real effort to enhance communication at all levels is under way in the RCMP through the use of electronic mail, newsletters, and a broader distribution of information that contributes to corporate transparency. Murray's (1998) Directional Statement emphasizes the importance of improving internal and external communication. The commissioner points to the importance of effective communication in addressing common issues and creating opportunities for continuous learning. Communication is the basis through which alignment and cultural change will occur.

Futuristic Management Concepts in Community Policing

A two-year demonstration project in Lethbridge and Grande Prairie, Alberta, completed in June 1994, focused on "developing a futuristic model for efficient and effective delivery of community-based policing services with an emphasis on delegating authority to result in increasing autonomy at the unit level" (RCMP, 1992). Input from all members was solicited to identify issues, policy, or procedures that prevented implementation of community policing. A total of 117 recommendations were received, and 71 recommendations were implemented.

Shared Leadership Vision

The Shared Leadership Vision process, undertaken in 1996, involved approximately 15 percent of all RCMP employees across Canada. The SLV workshops resulted in developing a set of guiding principles and a collective vision for future directions as Canada's national police service.

NATIONAL CRIME PREVENTION COUNCIL OF CANADA

In February 1993, the House of Commons Report "Crime Prevention in Canada: Toward a National Strategy," included a discussion of different approaches to crime prevention, measures to prevent crime, and the federal role in crime prevention. The first recommendation of the report was to develop a national crime prevention policy. The second recommendation was to form a national crime prevention council.

> The National Crime Prevention Council was founded in July of 1994. . . . Its mandate was to provide advice to governments on the design, delivery and evaluation of a comprehensive prevention strategy, and be a resource to communities who are involved in prevention initiatives (National Crime Prevention Council, 1997a).

The Second Report of the National Crime Prevention Council of Canada (NCPC) describes the focus and priorities of the council, the value of community-based problem solving, the rationale for investing in communities, and building real partners for prevention. The report makes clear the need to focus on developing goals for crime prevention, identifying the means to achieve them, and implementing the delivery strategy most likely to succeed. Using a community-based problem-solving model that is not unlike CAPRA, the council identifies five key participating groups—communities, the justice system (with an emphasis on the police), nongovernmental organizations or associations, the private sector and labor, and governments at all levels. These groups should build strategic partnerships toward a "comprehensive national strategy for community mobilization to prevent crime, ensuring that all communities have access to the knowledge, skills, and resources they need to get the job done" (National Crime Prevention Council, 1997a:7).

The NCPC recommends that key partners provide the resources, skills, and knowledge communities need to mobilize in order to prevent crime. The council also recommends that its primary role should shift from policy development to one of supporting the development of a national strategy for community mobilization to prevent crime (National Crime Prevention Council, 1997a).

The RCMP as Community Partner in a National Crime Prevention Strategy

The conclusions and recommendations of the NCPC represent timely leadership and the basis of a national crime prevention strategy. The conclusions appear to be appropriate and consistent with the current body of knowledge that recognizes the complex and multiple causes of crime and the equally complex and comprehensive response needed to prevent crime. The call to mobilize the key partner groups to take action toward achieving a national crime prevention strategy is one that should easily be heard by the potential partners as they are stakeholders in developing the most effective response possible. The RCMP must follow the leadership of the NCPC and respond to their call to mobilize for action.

Many examples illustrate the best practices of the RCMP as a key or strategic partner in the delivery of comprehensive, integrated community-based crime prevention services to children and their families. The following are representative examples.

Parkland Healthy Families—Stony Plain, Alberta

Parkland Healthy Families is an interagency, professionally initiated association with many member agencies. This nonprofit association administers the Brighter Futures Program, a program for families with high-risk children from birth to 6 years of age. The association also administers the Turning Points Program which is funded by Alberta Mental Health and provides services to individuals in abusive relationships. Both programs are built on ongoing needs assessment and evaluation. Member agencies include the RCMP, Victims Services, mental health agencies, medical association, schools, and many others (RCMP, 1998).

Pathways to Wellness—Beauval, Saskatchewan

Pathways to Wellness is a community-initiated process that aims to improve the lives and health of residents in the northern community of Beauval. This grassroots effort attempts to meet the goals of promoting and improving community relations, personal and community healing, sustainable economic development, youth development, community participation, cultural and spiritual development, and physical health and fitness. The project addresses a number of issues that directly affect children, including such topics as the need for support and education of parents, family violence, alcohol and drug abuse, and apathy among youth. Pathways to Wellness recently received Brighter Futures funding to hire a child development worker who will coordinate the development of a day-care center and parenting education and support (National Crime Prevention Council, 1997).

Community Justice/Family Group Conferencing—Lethbridge, Alberta

Based on an existing model in Sparwood, British Columbia, Community Justice/Family Group Conferencing training was provided to 70 stakeholders from across western Canada. Subsequent training has been provided to 150 community members, including approximately 50 police officers. The Family Group Conferencing is a precharge, problem-solving initiative that addresses criminal behavior and conduct or discipline issues involving children and criminal behavior, workplace or sexual harassment, or family violence involving persons over age 18. The objective is to divert eligible and recommended instances of conflict to a community-oriented problem-solving process. In this process, the victim and supporters have a voice in the outcome, the emotional impact of the conflict is acknowledged, the community is involved adding new problem-solving strategies in schools, the police role in the community is positively enhanced, offenders accept responsibility and agree to compensate victims, the problem is addressed quickly and responsibly, and court costs, lawyer fees, police expenses and witness expenses are reduced (Hickman, 1998; McDonald et al., 1995).

Future of RCMP Crime Prevention: Some History Is Not Worth Repeating

The future of RCMP crime prevention is at a crossroads. It is important to determine whether the existing programs are relevant, or whether they have outlived the purposes for which they were intended. It would be a mistake to eliminate crime prevention programs that are still relevant and useful in close-knit rural communities.

The future of RCMP crime prevention is situated within the community policing philosophy and service-delivery strategy, using CAPRA as a problem-solving tool. The RCMP must act as a catalyst to educate and mobilize communities into accepting their role as partners in solving community problems. It must strive to fully implement community policing and support and reward community problem solving. It is important to recognize the inherently strong elements of crime prevention present in the community problem-solving strategy, and promote its continued use. The RCMP must recognize the expertise and first-hand knowledge of front-line members, who should be encouraged to contribute to future developments in crime prevention planning.

In keeping with partnerships within the community and the alignment of service delivery with community policing philosophies, the RCMP must follow the leadership of the National Crime Prevention Council and support the council's recommendations by accepting its role as a strategic partner in the development of a national strategy for community mobilization to prevent crime.

ENDNOTES

1. An internal process to create guiding principles in the RCMP, approximately 18 percent of all employees contributed to the development of a shared mission, vision, and values statement, with a stated commitment to employees and communities.
2. RCMP Fact Sheet, No. 28, confirms the program orientation of RCMP crime prevention, and further states that the role of the branch is to focus on social order needs through education and public awareness strategies and the promotion of interagency cooperation and personal contact. The specific needs of crime victims are addressed through responsive community-based police services.
3. Ward (1997:3) states "the problem with community crime prevention programs is that they reflect America's reactionary society and its aversion to long-term planning . . . , [which] results in disjointed crime prevention activities." It may be time to assess the validity and effectiveness of the apparent program-oriented approach to crime prevention.
4. Ward (1997:3) observes that "block watch and/or mobile patrol programs work best in areas with few social problems and in communities that have already established networks for crime prevention." This situation applies to a great number of the communities the RCMP police across Canada.
5. "Society needs crime if it is to survive; without crimes, there would be no punishment rituals. The rules could not be ceremonially acted out and would decay in the public consciousness. The moral sentiments that are aroused when the members of society feel a common outrage against some heinous violation would no longer be felt. If a society went too long without crimes and punishments, its own bonds would fade away and the group would fall apart" (Collins, 1982:112).
6. In fairness to RCMP members working in the field of crime prevention, the recommendation to form a National Crime Prevention Council to develop a National Strategy for Crime Prevention was first made in February 1993 in a House of Commons Report (Horner, 1993).
7. Greene et al. (1994:93) advise that "the study of the adoption of community policing, then, is the study of organizational structure, culture, and service delivery change."

Chapter 4

The Crime Prevention Continuum: A Community Policing Perspective on Crime Prevention in Canada

John Lindsay

This chapter addresses one of the enduring, but erroneous, myths of modern policing: Crime prevention is nice, but cannot really work. What is less evident from this simple statement is that this mistaken belief belongs with other similar myths of prevention commonly attributed to policing. Principal among these is that routine patrol, rapid response, and after-the-fact investigation discourage the commission of crimes and criminal activity. What is masked by these characteristics of the crime suppression orientation of another policing era is that their real relationship to preventing crime is contingent upon random chance or good luck. Not many would risk the safety of neighborhoods, or their own professional reputations, on such tentative hopes for good fortune.

The paradox is that, as currently and generally practiced, the activity of crime prevention generates more public fear and insecurity than it does safety. This fact is borne out by the reality that, in both absolute and relative terms, during 1998 more crime was committed than in 1988, the fear of crime has increased, and policing costs are rising. These are the unintended consequences of an imperfect crime prevention design, which is actually inconsistent with the other understandings of modern policing.

Canadians have always desired and defined Canadian society by the need for peace, order, and good government. This, therefore, must be the intended consequence of Canadian policing. This chapter will strongly suggest that all of this is achievable within a crime prevention continuum that commences with crime suppression, moves to prevention programming, and concludes with community building.

The opportunities that can deal more realistically and inclusively with the status quo to reduce social inequality in prevention activities are exciting! The quality of life for all citizens can be very much enhanced by a more holistic approach to, and understanding of, crime and disorder reduction strategies. The key is to recognize the inherent mutual dependence between community-based policing and crime prevention, and to achieve it. The goal is to understand that the police role—wherever on the continuum a particular issue belongs—is to stop bad things from happening.

POLICING IN CANADA

Canada is an enormous country that much of the world still envisions as a rugged and empty wilderness populated by equally rugged and self-reliant individualists. While this characterization may have been quite accurate not so long ago, the contemporary reality is that Canada is a complex and largely urban, cosmopolitan, and post-industrial society. However, the psyche of even "modern Canadians" is still shaped by its physical geography and the lure of a young and empty land. For these reasons, the analogy of a "white water journey" upon a rushing and dangerous river is attractive to Canadians as a symbolic representation of the challenging route that police and community must travel together to achieve a shared vision of community-based policing.

To an audience of police professionals, a journey of a different kind may be much more appropriate—a journey in time rather than distance. The history and state of Canada's policing is a story of different times rooted in a very recent past and held together in sequence by the golden threads of public service and the rule of law. One can characterize this history of Canadian policing by playing on a popular movie title and offering a journey "back to the future" as the best understanding of "Canadian policing." By re-connecting with the past, it is suggested that one can discern the future direction in policing Canada as a simple journey along the crime prevention continuum.

Back to the Future

Canada is a young country and its history of formal policing is recent. It is perhaps the freshness of pioneering times in the collective memory that favors Edmonton with an intuitive understanding of community policing. It is to state the obvious by declaring that community policing is something old that is new again. While Britain provided a framework for the development of policing in Canada, history and geography may have led Canadians to learn a different lesson.

Less than 150 years ago, western Canada was untamed frontier consisting of vast prairies and forests cut by thousands of rivers and backed against the Rocky Mountains. These Northwest Territories were controlled by trading companies and their interests were foremost in this raw land. The power of furs, liquor, and firearms were used with great effectiveness to govern for profit.

The government of Canada acquired these territories in 1870 and began to encourage their settlement by new immigrants. The offer of free land brought people from many parts of the world. Within three years, the Canadian Parliament passed an act to provide for the formation of The Northwest Mounted Police. A year later, a force of 275 men made an epic march from the most western outpost of civilized Canada, in what is today Manitoba, to the foothills of the Rocky Mountains. The "law" had arrived, but not the law as it is known today.

From the beginning, the police were at a disadvantage in manpower and munitions. Rule by force was not an option for these pioneer policemen. The police, often perilously outnumbered, would venture into hostile territory to effect an arrest for criminal or rebellious acts with remarkably few casualties on either side. A hard-earned reputation for temperance and fairness in the application of the law carried over to more peaceful times. Lone officers would patrol large areas, representing order and government as one of the few, and often only, visible symbols of nationhood. The RCMP were pragmatic veterans of the frontier with keen survival instincts. They recognized a fundamental principle of survival—that of mutual dependency.

These policemen were dependent on the inhabitants of the land. Food, shelter, and fresh horses were asked for and received with the mutual recognition that the pioneer and police were essential to each other's survival. The police learned to respect those they depended upon, regardless of race or social status. To the settlers, the police offered a visible reassurance of the existence of order and good government. They mediated civil disputes, investigated criminal matters, and generally attempted to keep the avarice of the gold rush and whiskey traders at bay.

The concept of mutual dependency carried on as farms, towns, and cities began to spread across the landscape. Because of the isolation and distance between settlements, pioneering families depended on their neighbors. Whether at calving time or the birth of a child, the unwritten code between neighbors was the duty to help. When community safety was threatened by natural or man-made disasters, the police officer could expect the local residents to answer his call for help. Quite simply, it was through a bond of mutual dependency that the police were able to secure the willing cooperation of the public.

Of course, other factors besides geography and a hostile environment facilitated the appreciation for police/community partnerships. Pioneer western Canadian society lacked the degree of social stratification and class structure that had developed over time elsewhere. This allowed a sense of egalitarian optimism that gave everyone hope for the future. In an environment of expansion, peacemaking allowed more freedom of opportunity than what may have been tolerated in an older, established society more concerned with order maintenance.

Professional Policing

Turning to present history, Edmonton, a typical Canadian city, is struggling to disentangle itself from the professionalization of policing that occurred after the first World War. In the United States, a growing intolerance for government and police corruption led to a call between the World Wars for the separation of police and public. The development of professional standards of behavior, accompanied by a greater reliance on scientific methodology and due process of law, were the result of the growth of liberalism. Self-determination and individual rights became paramount values within western societies. Canada followed the U.S. down the road of police professionalism and, in turn, this new police culture had a powerful influence on both the police and public. Technology supported a new industry that fit perfectly with the expectations of a new breed of young, educated police officers. The pervasive effect of this self-imposed professional isolation was to marginalize the community, and therefore to retreat from the embrace of mutual dependency.

The modern quest for professionalism was a necessary step and many advantages were gained. Equally important, however, is to acknowledge what was lost. As a result of the abandonment of the concept of mutual dependence, the police lost mutual respect with the public. The police began to lose their legitimacy as peacemakers and, in corresponding degrees, increased their reliance on the use and threat of force to achieve public policy. Urbanization and the growth of the welfare state also brought a general deterioration of the duties of citizens to help each other in achieving community peace. The professional police encouraged this by claiming ownership of law and order, externalizing responsibility for community values from the community itself.

With a renewed Canadian commitment to community policing, police leaders are now seeking to reconnect with the past and again discover the meaning of mutual dependence. This is, of course, both the key to effective policing and crime prevention, and the real reason why crime prevention works.

THE CRIME PREVENTION INDUSTRY

Ironically, crime prevention represents a real "threat" to community policing. Mutual dependence does not yet characterize the relationship of these two strategies. While they should be complementary, they are too frequently found in conflict, fighting for ideological supremacy and scarce resources even within the same organization. Their present utilization as opposites represents a cultural schism in policing and renders the conflict predictable, as well as inevitable. At the risk of overstatement, while community policing is essentially inclusive and holistic, traditional program-oriented crime prevention is too often applied in a "them vs. us" context where "us" is defined not as police but as those members of society that reflect the dominant and advantaged culture, and "them" is everyone else. This apparent division of community by the police is the very antithesis of community-based policing.

Of course, the practitioners of this style of crime prevention do their work in good faith and rarely consider themselves representatives of the dominant culture. In North America, the normative standard for that "dominant culture" is the white, English-speaking male from the higher socioeconomic strata who holds a set of political and religious beliefs falling within right-centrist parameters. Those outside of this spectrum can too easily become symbols of mistrust and uncertainty, with these perceptions increasing in intensity the farther away they are perceived from the embrace of convention. The status quo is rarely egalitarian and does tend to serve the expectation that more protection and more safety is required for, and by, vested interests.

Within the current policing lexicon are found phrases such as "community safety" and "safe cities", which tend to present the implied question to an anxious public: safe from what? It is too easy to respond to that question by inferring that the nonnormative elements of society are deservedly targeted by traditional crime prevention. The poor, socially misanthropic, mentally disturbed, and culturally foreign bear the stigma (not supported in reality) that they may threaten the safety of the community. This lack of trust and knowledge are the elements that generate fear and make certain groups targets of crime prevention strategies and overpolicing. This is a social commentary that requires both acknowledgment and correction in the understanding of a new crime prevention continuum.

Another classic crime prevention response has been to support those who have something which is vulnerable to loss. Whether extrinsic, such as material possessions, or intrinsic, as with social status and personal security, those perceived valuables are resident with the "haves" and, it is assumed, coveted by the "have-nots." A crime prevention industry has formed, consisting of opportunists within both the public and private sectors, that is willing to provide information to the point of overload on the dangers the average citizen faces every day. This activity is followed by costly programs, services, and gadgets to both soothe and reassure the frightened public psyche. While this process has common and conventional understanding, and may even be logical in a purely rational way, it cannot be the best expression of crime prevention. Some of the reasons should be self-evident from two examples of traditional programs that inevitably have had unintended consequences.

First, much of the contemporary Canadian prevention program orientation is directed towards children, advising them to be wary of strangers, to be fearful of traffic, and to be hypervigilant to the potential dangers that surround them. Is it merely a coincidence that too many playgrounds stand virtually empty or unused, youngsters no longer walk to and from school, and that their permissible social contacts are increasingly confined to an ever-smaller cadre of screened and approved peers? Modern society has assigned a name for this phenomenon—it is called "cocooning." Though rarely acknowledged, this inward and introspective focus is facilitated by safety messages which discourage the very neighborhood activism required for community policing.

Second, crime prevention professionals can readily advise people of the many ways a house is not safe to live in, or the streets not safe to walk on. The very young, the very old, and women of all ages are constantly advised that they are particularly vulnerable and should consciously restrict their movements so as to maximize their opportunities to remain safe. The unintended consequence is that there is an increase in the fear of crime or victimization within these target groups. At the same time, a public bias against youth and the poor as potential or probable offenders is reinforced by a crime prevention industry that is, by result, fear producing.

The results, therefore, are not surprising. Even though recent trends show a decline in crime rates, successive polls have indicated that fully 50 percent of Canadians feel less safe today than they did five years ago and nearly the same number feel that violent crime is on the increase. In fact, after unemployment, Canadians believe that crime is the most important Canadian social issue. The fear of crime has particular affect upon women, with 48 percent indicating they are afraid to walk alone at night in their own neighborhood, versus 18 percent of men (Gallup, 1993, 1995; Reid, 1990).

Unfortunately, the crime prevention strategies of the past two decades have not worked. Despite a fourth consecutive year of marginally decreasing crime rates (down 1 percent in 1995), the overall crime rate in Canada is still 8 percent greater than 10 years ago. Violent crime has increased fourfold since the 1960s and violent youth crime is growing at twice the rate of adults, having more than doubled since 1986 (Brantingham and Eaton, 1996).

Within this growth scenario is the realization that a large number of crimes also go unreported. One survey of Canadian women found that only 11 percent of sexual assaults were reported. Similarly, police were advised of only 26 percent of spousal violence cases. The 1993 General Social Survey recorded that an astounding 72 percent of violent victimization went unreported (Canadian Centre for Justice Statistics, 1994). The only general message, therefore, to be drawn from these disturbing findings is that traditional crime prevention is in disarray and that this may be affecting the credibility of the police.

CRIME PREVENTION CONTINUUM

Many policing professionals view crime prevention with capital letters. It is more like a placid noun than an action verb. Through their eyes "crime prevention" is an office that delivers programs to people who are free to take it or leave it. Either way, the reactive model of policing will be there to answer the crime calls and will, almost without exception, invest much greater resources in crisis response than prevention.

On the other hand, crime prevention, in its elemental sense, is about stopping bad things from happening. That is the core function of policing. Community policing accommodates this understanding and represents the best approach with which to accomplish the objective of crime prevention. Through community policing, the tools of crime prevention can be applied by police agencies in progressive order toward enhancing a community's ability to protect itself from unhealthy elements, whether criminal or social.

Expressed as a crime prevention or policing continuum, the tools of crime prevention available to police agencies are (1) crime suppression, progressing to (2) prevention programming, and developing into (3) community building. These three strategies share the crime prevention continuum, and the effectiveness of their application can be expressed as short-, medium- and long-term, respectively. Whether on a micro- or macroscale, police response to all crime and disorder problems can be positioned somewhere on this continuum. Understanding and implementing the progressive stages of the continuum is key to achieving true crime prevention.

Presented below are some of the best practices of the Edmonton Police Service (EPS) within each stage of the continuum. These examples are typical of Canadian municipalities. The term "community," as used within this chapter, can be a geographical region with physical boundaries or an association of individuals sharing a common interest.

Crime Suppression

Crime suppression can be defined as the application of physical and legal resources of sufficient quantity and duration to apprehend, displace, or discourage the visible representations of crime and disorder. As the first stage of prevention, crime suppression seeks to reassert the exercise of legitimate authority and return control of public spaces to the community. This strategy may be required whenever the state of disorder has reached such a level of acceptance or acquiescence that the capacity for community action is not present.

Serious Offender Program

Recognizing that less than 5 percent of the population creates 80 percent of the criminal activity, the Serious Offender Program works in conjunction with the Justice Department to identify habitual and serious offenders who demonstrate a persistent lifestyle of criminal conduct. A points system is used to ensure uniformity in the application of this program. Points are accumulated by the offender, based on the severity of crimes committed, and these points accrue until a threshold is reached and he is formally identified as a serious offender. Subsequently each new serious offender is assigned to a detective within the program. The single objective is to keep the offender "off the street," either through rearrest for new offenses or for breaches of release conditions or by attending bail and parole hearings to advocate for continued incarceration. Additionally, greater use of Dangerous Offender legislation to obtain open-ended prison sentences for the most serious habitual offenders is being pursued by justice officials with greater frequency and tenacity. Though pursued through reactive suppression tactics, such work is a direct manifestation of community needs.

Hot Wheels

The Hot Wheels Program illustrates a proactive application of suppression tactics. Crime reports identify areas where auto thefts occur with great frequency, such as shopping centers. Those areas are reconnoitered until one or more stolen autos are located that appear to have been "laid down" for further use. Surveillance is maintained on those vehicles, and when the culprits return and attempt to drive the stolen auto away, police officers move in to effect the arrest of the driver and any passengers. This single tactic has contributed to a 6 percent decrease in auto thefts in Edmonton, despite regional and national trends to the contrary.

Prevention Programming

Prevention programming represents more traditional crime prevention activity. In this stage, service providers, including the police, deliver prevention programs to individuals and communities who are thereby motivated to implement the programs through self-interest. These programs generally focus on the reduction of property and personal crimes through changing the behaviors and/or the environment of potential victims. This is a developmental stage which demonstrates to communities that they can act together in mutual dependency to influence crime.

Like most Canadian municipalities, the EPS sponsors the activities of citywide Neighborhood Watch and Block Parent associations. Each program is provided with office space and administrative support, and in return these associations provide full-time staff and a volunteer organization dedicated to prevention programming. Their volunteers organize neighborhoods into telephone crime alert networks, provide home safety audits and deliver child safety lectures to elementary school children. They also provide free videotapes and coloring books to communicate their message of crime prevention and child safety. These two groups represent a pool of over 10,000 volunteers, which the EPS can call upon when required.

On a smaller scale, prevention programming is also provided by the officers of several specialized investigation units within the EPS. A member of each detective unit develops lesson plans and training materials to present to at-risk groups. For example, the Child Abuse Unit recently completed development, along with the police audiovisual department, of a videotape for single mothers on how to recognize a sexual predator. Another video was created for use in joint police–child welfare training on the psychology of sex offenders. These are just two examples of the type of educational awareness that can be raised by the police concerning "hard" crimes.

It must be noted that prevention initiatives need not be limited to criminal activity. In 1997, the Truck Detail determined that the mechanical state of school buses within the city had deteriorated to a substandard, even dangerous level. The prior deregulation of government inspections had left the responsibility for safety checks with the school bus owners and the private operators. Fortunately, however, the inspection and enforcement powers of the police under transportation laws were unaffected.

During a yearlong effort, which drew heavy criticism from the private school bus operators and government, hundreds of mechanical inspections were conducted by specially trained officers at schools throughout the city. At the beginning of this inspection period, as many as eight of every ten buses failed to such an extent that they were not allowed to drive away. Notwithstanding continued media pressure, the public remained favorably disposed to these police inspections, which continued. By 1998, major purchases of new buses had been made by the bus operators, and the government had increased their funding for student transportation to a more realistic level. The only reasonable conclusion to be drawn from this experience is that the norm within that community of interest (school bus owners and operators) has been elevated to the extent that nine of ten buses now pass the prescribed inspection.

These multiple efforts at prevention tend to be cumulative in their effect in developing a community's capacity to move forward in the continuum. That is always the goal.

Community Building

Community building can be equated with social development in the sociological milieu, but for policing purposes it is the ultimate stage of prevention where communities have recognized that the root causes of crime must be addressed. Drawing on what is learned from the previous two stages of crime suppression and prevention programming, communities develop insight on societal causes of crime and act with a long-term view to diminish them. A community can do no better than this.

In October 1996, the National Crime Prevention Council of Canada brought together leading researchers to discuss crime prevention through social development. This group identified eight social strategies that, in isolation or together, could form the basis of an effective crime prevention strategy (Canada, 1997). Some examples of these "made in Canada" solutions can be derived from the EPS where interdisciplinary work with many partners has long characterized a commitment to crime prevention through social development.

Early Education for At-Risk Preschool Children

Success by 6. Study groups have demonstrated that at-risk children benefit from early education and are ultimately five times less likely to engage in criminal behavior than those who do not have the opportunity to take part. Success by 6 is both a program and a concept that promotes collaborative involvement from all sectors of the community, including the police, toward the common purpose of helping children succeed for life.

The goal is for children to reach the age of 6 having been nourished, both mentally and physically, and having learned the rudimentary basic skills of conflict resolution. All this is achieved directly with the child and frequently with her primary caregiver through education in parenting skills. When achieved by 6, such nourishment results in a marked diminishment in the chances of becoming a criminal or delinquent by age 14!

Success by 6 was pioneered by the United Way of Minneapolis in 1989 and, since then, approximately 28 communities across North America have initiated similar efforts. Edmonton is the first Canadian city to launch Success by 6, with full-time police personnel assigned to work with the team. The project's mission is to facilitate collaboration among government, nonprofit groups, business, and communities to strengthen preschool children and their families through prevention and integrated service delivery in health, education, and social services.

Protects. Though a less intensive early childhood education initiative, PROTECTS is unique. Under this banner, tactical team members visit classrooms for the single purpose of reading stories, giving life to the acronym

PROTECTS—Police Reading Outloud to Educate Children through Stories. The program targets inner-city children at the kindergarten level. At-risk children are selected for this one-on-one experience and are read to by the same police officer once a week for a 15-minute period. In addition to the obvious benefits of increased literacy skills, this initiative also has proved very successful in giving children a different and early positive perspective of police. The secondary objective of providing a real and positive role model for how "powerful people" should behave is extremely beneficial, and directly transferable to the child's relationships with others "less powerful" than themselves.

Social Skills Groups for Parents and High-Risk Children

Dare. Drug Abuse Resistance Education (DARE) is a comprehensive prevention program designed to equip grade 6 students with the skills to recognize and resist social pressures to experiment with cigarettes, alcohol, and other drugs. This unique life-skills program was developed in 1983 by the Los Angeles Unified School District and utilizes police officers to teach a formal curriculum in the classroom. The choice to target grade 6 students was deliberately made to prepare them for entry into junior high school where they are most likely to encounter the temptation, opportunity, or pressure to use drugs.

A recent enhancement to the program has been the inclusion of antiviolence, antigang messages in the curriculum. Parental involvement is also a significant feature of the program. Through fund-raising activities, displays, and presentations, parents are encouraged to support the lessons of DARE in the home as well as in the school.

The wide acceptance of DARE as a beneficial exercise of police action has been fully legitimized in the Canadian context by the establishment of the DARE Canada training school within the EPS. Fifty-seven schools participate in the DARE program in Edmonton, and another 92 are on a waiting list. A full evaluation of this initiative is also a work-in-progress at the University of Alberta.

Pride through Play. Pride through Play is a unique project developed in response to numerous studies in Edmonton which revealed a lack of recreational, leisure, and social programs for youth and families in low-income areas. Coordinated by the city's Community Services Department, and with active participation by the EPS, the project seeks to develop youth leaders who will initiate recreational activities. These activities must be designed to encourage the development of good social skills, promote children's need for play, or help young people and families gain self-esteem and confidence.

Home Visits to Reduce Child Abuse and Violent Behavior

Partners for Kids and Youth. Through participation in Partners for Kids and Youth, a multiagency program, police officers are connected with agencies that provide outreach services in vulnerable communities. In this program, the vulnerable community consists of children who are, or may become, abused. These outreach services include public health nurses, social workers, and child psychologists. When these partners determine a need for police assistance or intervention, the officers accompany them on home visits in a nonthreatening manner to build the necessary trust to implement effective solutions.

Youth Options. Youth Options is a similar collaborative effort involving support agencies. This project connects young people with the community by encouraging them to identify their concerns and find solutions with community support. By providing young people with options in their community, increased participation of teenagers in community activities is achieved. To date, those participating in the program have created a movie night where donated videos are shown at the local recreation center, they have entered a team to represent the City of Edmonton in the North American Indigenous Games, and they have started a lunch-hour weight lifting program at a local school. In this latter case, the activity has eliminated calls for service along the local thoroughfare during lunch-time hours.

Community-Based, Multidimensional Approaches to Intervention

CARRT. The Child at Risk Response Teams (CARRT) are a joint initiative of child welfare authorities and thc EPS. Four teams, each consisting of a child welfare worker and police constable, are available for dispatch to initial complaints of children who are at risk of neglect or abuse. The purpose is to provide the best possible initial response to suspected cases of abuse. They ensure that at-risk children are identified early so that referrals can be made to appropriate services for more effective intervention. During 1997, four permanent teams have dealt with almost 1,500 children on a level of competence far in excess of that expected of a general response unit.

Community Conferencing. Hearing the never-ending frustrations with the legal system of victims and communities, the EPS undertook to develop a restorative justice alternative by adopting a Family Group Conferencing model from New Zealand. Renamed Community Conferencing, offenders are brought together with victims and others affected by their crime in a controlled environment where all parties consent to participate. A solution agreeable to all is sought, which restores the victim's sense of security and provides meaningful shaming and consequences to the offender. The family and friends present from each side become part of the supervision and support plans for both offender and victim.

A period of evaluation focusing on community conferencing and youth crime proved to be highly successful, resulting in the establishment of a full-time community conferencing coordinator and a dozen trained facilitators, within the EPS. The EPS has partnered this initiative with the local John Howard Society, a prisoners' rights group, to jointly provide this restorative justice alternative in appropriate cases. This approach was designed with assistance from Canadian native Indian communities and is a welcome example of aboriginal values influencing the dominant culture. Great success with this model of restorative justice has also been achieved by the Royal Canadian Mounted Police in many communities across the country.

Helping Children Who Witness Violence

Through the implementation of the Child Witness Court Preparation Program, children are led through various games and exercises designed to lessen their fear of giving courtroom testimony, and the impact of the abuse or violence to which they have been exposed. The program identifies children who require additional support or psychological counseling, and it seeks to return the children and their families to normalcy as soon as possible. In this fashion, it is much more than just a support system for the Crown prosecutors.

Reducing Social Inequality and Child Poverty

When local governments eliminated funding for kindergarten classes, many police agencies and other interests publicly criticized the move as placing disadvantaged children at greater risk of abuse and neglect. The funding was eventually restored. The dissemination of police experience with such situations may be regarded by some as social activism, but it is an important component of being "committed to community needs" (part of the EPS core values). Crime prevention through social development requires much more than traditional police skills or advocacy.

Reducing Access to Guns

In this arena, the EPS is optimistically moving forward. The federal government has major legislation scheduled for implementation this year that will eventually result in the registration of all firearms. Storage requirements and firearms education are also part of the legislative package. However, since the provincial government has opted out of implementing the federal law and police agencies, the community must be patient in developing solutions to utilize the worthwhile objectives set for this legislation.

Limiting Alcohol Use

Society has seen the successful application of community action to effect a drastic reduction in the acceptance of alcohol abuse. Organizations such as Mothers Against Drunk Driving (MADD), Students Against Drunk Driving (SADD), and Parents Against Impaired Driving (PAID) have effectively communicated that message, prompting increased enforcement, improved education, and new legislation allowing for vehicle impoundment and mandatory jail sentences for second offenses. Without such support the police could never have made such a positive impact. These initiatives have, in turn, developed other highly effective detection and deterrent programs (for example, Operation Lookout, Operation Red Nose).

WHAT DOESN'T WORK

This brief synopsis of prevention strategies reveals how police, professionals, and communities can act together to effect solutions. The greater level of detail required to describe community building activities reflects their greater effort and potential. It is important, however, to caution those who seek answers through transplanting successful strategies from outside their own community. Although "best practices" often reveal innovative thinking, the learning that takes place during the design and implementation of local solutions has many more benefits than a single outcome to an individual problem. Each time success is achieved, the community is stronger and better able to tackle its next challenge. To achieve learning within the crime prevention continuum, the process is as important as the outcome.

During the learning process, the EPS discovered many unworkable tactics, and these bear mentioning as much as those which have worked well. There is always a need to analyze failures as thoughtfully as successes, if only to avoid needlessly delaying continued progress.

First, within the Hot Wheels Program description, it was noted that Edmonton had experienced a decrease in auto thefts even though surrounding areas had experienced an increase. In fact, neighboring municipalities experienced a substantial 30 percent increase, which is likely an example of the displacement effect of suppression tactics. Prior notice of anticipated displacement, therefore, should be provided to adjoining jurisdictions so they may plan an appropriate local response.

Second, formal crime prevention programming within the EPS has not been required to conform to the community policing expectations that were imposed upon other areas of the service. Although the concepts of ownership and problem solving have been applied by front-line members, crime prevention was handled by "isolated specialists," with an exclusive focus on general education and no requirement to directly support street-level policing. The realization is emerging that crime prevention practitioners could be real problem-solving specialists if their objectives and strategies were redesigned within a community policing context to achieve this outcome.

Third, community policing activities have achieved meaningful consultation at the grassroots level, with important work being accomplished through short-term or tactical problem-solving efforts. The drawback is the lack of a meaningful consultation process at a level above the local neighborhood. The broader conclusion is that to achieve citywide outcomes, police agencies must mobilize across communities and adapt broad consultation strategies. To remedy this situation in Edmonton, general public forums will share information and solicit feedback on the broader issues of policing. Focus groups within these forums will discuss such topics as "police and youth" and "police and communities." Just as important as the immediate feedback, networks will be established upon which to build greater community participation in police decision making.

POLICING IN THE CRIME PREVENTION CONTINUUM

The police role in the crime prevention continuum is unique. Police officers are daily asked to join community prevention initiatives large and small. In many cases, it is the police officer who is the recognized expert on community building, who guides the planning toward long-term sustainable solutions, and who is asked to represent the conscience of the community. A truly humbling responsibility has been placed in their hands.

Other professionals and communities at large have lost their understanding of mutual dependency. A renewed commitment to this fundamental social value is essential to forming the partnerships required for crime prevention through social development or community building. Police agencies, therefore, need to accept the role of implementing the crime prevention continuum, but only by passing on their knowledge and sharing the responsibility for crime prevention.

Police can determine where on the continuum a particular issue belongs, based on the extent of the problem and an assessment of the community's ability to respond. Through a community policing style and a problem-oriented approach, police agencies can implement strategies within each stage of the continuum, either by suppressive means in the early stages or through education, partnerships, and community empowerment in the latter stages. In addition to implementing the appropriate response, police agencies must assume an evaluative role to guide their efforts and ensure a progression along the continuum. Clearly, when a prevention initiative has been demonstrated to be self-sufficient, police practitioners must then set up a maintenance mechanism to alert when—not if—further attention is required. Sustaining a program is always as important as its creation.

In certain circumstances is it possible for a community or a problem not to progress beyond suppression or prevention programming? Certainly. Areas with very low social cohesion present special challenges in this regard. For this reason it is always important to evaluate the degree to which the needed social structure is being developed. For example, gang activity represents social structure just as much as a school Parent-Teacher Association. Where one exists, so can the other. The challenge, and the objective of community building, is to encourage a legitimate alternative to destructive versions of social organization.

Through community policing, citizens are learning that, together with the police, they can optimize the role of crime prevention and build healthy communities. They really can stop bad things from happening!

CHAPTER 5

CRIME PREVENTION IN THE NETHERLANDS: A COMMUNITY POLICING APPROACH

Alexis A. Aronowitz

With the enactment of the 1993 Police Act in January 1994, the Dutch police underwent a major reorganization from one national and 148 municipal police agencies[1] (a decentralized model) to 25 regional[2] and the National Police Services Agency (a more centralized model). The rationale behind the restructuring was to improve the police organization, which at the time was fragmented, inefficient, expensive (Van Reenen, 1993), and incapable of coping with the changing crime climate. Within the 25 regions, the newly decentralized structure reduced the number of hierarchical levels within the organization and generalization of tasks (Horn, 1993). An important by-product of the increased efficiency and effectiveness of the police was an improved relationship with the public.

The development of community policing in The Netherlands followed the philosophy guiding the movement from a reactive, crime-fighting organization to a problem-oriented, proactive one. Early experiments with community policing placed officers on the streets with no clearly defined role and raised the question of the effectiveness of such a measure. Community-oriented policing gained new impetus as the government's emphasis on a "total approach to safety" and cooperative problem solving at the local level resulted in the police becoming partners with citizens, social welfare, and community agencies in dealing with the problems surrounding security and peoples' fear of crime. Realizing that citizens are the ones most closely affected by the problems and that they are the ones best able to identify concerns and possible solutions, many police-initiated prevention projects involve citizen input as well as cooperative relations with other social and justice agencies. Crime prevention projects in The Netherlands, therefore, can be characterized as varied, local, problem-oriented, and multiagency approaches.

HISTORY, PHILOSOPHY AND CONCEPT OF CRIME PREVENTION IN THE NETHERLANDS

Against a backdrop of skyrocketing crime rates (an almost 10-fold increase from 130,000 registered crimes in 1960 to almost 1 million in 1984), a commission was formed to study the problem and make recommendations to the government. The Commission Roethof drew a line between frequently committed, less weighty crimes and crime of a more serious nature. Although the latter should be the focus of police and justice agencies, less serious forms of crime, seen as social problems, could best be dealt with in terms of cooperative working relationships between such organizations as schools, businesses, citizens' groups, social agencies, and the traditional justice agencies and police. In terms of its focus on crime prevention, the commission recommended

- The mobilization of neighborhood committees or involvement of residents in the identification of problematic situations (so-called neighborhood prevention projects)
- The appointment of security guards in crime-ridden apartment buildings
- Strengthening the bonds of weak students to their schools or by involving them in extracurricular activities
- A more rapid repair of objects destroyed or damaged by graffiti, a more rapid demolition of houses deemed "unlivable" and abandoned business buildings, and the cleaning and refurbishment of dilapidated areas of the city in cooperation with the residents (Ministerie van Justitie, 1986:36)

The recommendations made by the parliamentary commission became the cornerstone for the national policy presented by the Minister of Justice to the Second Chamber of Parliament in 1985. The government recognized that enforcement of the law and crime prevention could no longer be the sole mandate of police and justice organizations. The policy should focus on differentiation and consistency. Differentiation implies that frequently committed less serious crimes should be dealt with differently than graver crimes. Less serious crimes should be fought with a combination of repressive and preventive measures. Consistency suggests that preventive measures be fine-tuned with the repressive measures of police and the department of public prosecution (Terlouw, 1991). The policy plan emphasized an administrative approach to crime prevention and recommended a cooperative approach between government, social welfare and health organizations, citizens, the police, and justice agencies.

On July 12, 1995, the central government entered into a formal agreement with the municipal authorities of Amsterdam, Rotterdam, the Hague, and Utrecht (these four cities came to be known as the G-4), establishing the groundwork for a long-term integrated approach to improving the quality of life and reducing unsafe conditions in these cities. Subsequent to this covenant, the central government also entered into a similar agreement with 15 other "large" cities which came to be known as the G-15 (and then with 6 others which came to be known as the G-6) (Ministerie van Justitie, 1997). This new plan, the Major Cities Policy, aimed at providing an integrated, neighborhood-based approach to problems, both in terms of the cooperative relationship between the central and municipal governments, as well as between judicial and social agencies, businesses, neighborhood groups and individuals within the cities.[3]

In recognition of the fact that justice organizations are often reactive, intervening after the fact, the Ministry of Justice introduced in 1997 a project entitled Stimulating Crime Prevention Justice. Its purpose was to assist justice organizations in developing crime prevention policy. The main focus of this project was to combine and disseminate information, to identify factors predicting success and failure, and to provide support in creating plans and a more accurate picture of certain types of crime.

One of the most important ideas to stimulate the development of and recognition of crime prevention programs in The Netherlands was the inauguration of an annual award presented by the Hein Roethof Award Commission. Founded in 1987, a substantial monetary prize has been awarded annually for the best crime prevention project in The Netherlands. In 1994, Belgium joined the competition (see, Hoefnagels and van Erpicum, 1997), followed in 1997 by the United Kingdom. There is now a European Prevention Award for nominations from the United Kingdom, Belgium, and The Netherlands (Ministerie van Justitie, 1997).

Community Policing

Because the police often serve as the first point of contact within the community when citizens signal an alarm, special attention should be given to the development of community policing. During the 1950s and 1960s, emphasis in policing was upon law enforcement and crime fighting. "Preventive" patrol took place in cars, and the police could best be described as a reactive force. The role of investigation was expanded in the 1960s and 1970s (Horn, 1993). Many of those neighborhood stations were closed during the 1970s as a result of the move toward more centralization (Horn and Koolhaas, 1990).

With the publication of the report "Police in Transition" in 1977, the Project Group Organizational Structures came to the conclusion that the police had not succeeded in adapting to the social developments of the 1960s and 1970s. The Project Group recommended that all police services be provided by small geographically decentralized units that are given a relatively large degree of independence. Furthermore, the police should strive toward integration into, and a good working relationship with, the community (Horn, 1993). The recommendations by this group led to the introduction of the neighborhood team model in some police forces in The Netherlands (Aronowitz, 1997b).

Ten years after the Project Group introduced its report, the neighborhood-oriented team policing model was fairly well integrated into a number of cities in The Netherlands, although not without some problems. The neighborhood teams varied according to the local situation, and ranged widely in size and function. Problem analyses were being conducted, although the quality of each analysis varied between neighborhood stations (Horn and Koolhaas, 1990).

Despite the fact that community officers had been successful at their attempts to become integrated into the communities they policed, they were not completely successful at ensuring order and safety in the neighborhoods. This was due, in part, to the fact that the first-generation police officers had been sent into the community to function as a sort of watchman with no clear-cut idea as to what exactly was their task. No specific policy guided their actions, and no job description was given. The position drew limited appeal from police officers, and the community officer gained little respect from his or her peers, often being given the name "neighborhood nurse" (Brinkman, 1997).

Neighborhood team policing was not a panacea for solving crime problems in the city. Neighborhood officers were often seen as the "public-friendly representative of the police." Addressing the wishes and demands of residents in the neighborhood often impeded the ability to initiate a problem-oriented proactive style of policing. A neighborhood-based approach did little to address problems that involved crimes that extended beyond specific neighborhoods, such as environmental offenses, fraud, and organized crime (Horn and Koolhaas, 1990).

Departments began losing interest in community policing when the focus shifted to organized crime in the beginning of the 1990s. Attention was directed toward the development of specialized and predominantly centralized units (divisions) of detectives. Additionally, a major reorganization of the police and a redistribution of police staff has led to the closing of small neighborhood stations (Gunther Moor and Van der Vijver, 1997).

The Dutch police, along with other municipal and justice agencies, were given an impetus in the form of the Major Cities Policy in July 1995. Districts were required to develop security plans, and an emphasis was placed on integral safety and cooperative working relationships between partners involved in security. The police became one of the focal points in local safety policy (Aronowitz, 1998).

More recently, community policing has once again become prevalent, bolstered by the fact that national and local politicians have demanded more police officers on the streets to reduce the public's sense of insecurity in their neighborhoods. Although community policing is making a comeback in a number of neighborhoods and cities, the nature of the task, the job description, and the training of community police officers has undergone radical changes. Whereas "crime prevention" among the police previously consisted of little more than handing out information brochures on ways to prevent crime, it has now become a movement within police forces and is seen as an important aspect of policing. Many forces now have active divisions that provide technical prevention advice by going into communities and inspecting residents' homes to inspect door locks and offer other antiburglary measures (Van der Vijver, 1998). In 1995, the National Center for Security through the Quality of Life was established. This center trains police officers and community workers responsible for performing work in the neighborhoods. An integral approach is the new philosophy guiding the Dutch police in which police officers work together with relevant partners in handling social problems in the local area. The neighborhood agent has become a liaison officer between the public, social work organizations, and justice agencies.

In 1997, the Police Inspectorate at the Ministry of the Interior conducted research into what has become known as "area-bound policing," translated as community policing. Recognizing that the police had to provide a qualitatively better product, in terms of accessibility and availability,[4] numerous police forces have once again begun concentrating efforts on community policing. Community policing in The Netherlands resembles forms of community policing in the United States and Great Britain. Characteristics of this style in The Netherlands are small-scale improvements in safety and the quality of life in general. Its focus is problem-oriented and network-, coalition-, and partnership-"minded."

The goals of community policing are to increase safety and the quality of life, on a subjective as well as on an objective level. What is necessary "to know and be known is realized through the means of a small-scale organizational form at the district-, village-, neighborhood-, or even street-level" (Inspectie Politie, 1997:15). There is one police agent for a few thousand residents (although this may vary). Geographical assignment of agents is permanent, in order to increase the recognizability and approachability of agents, and their only task is to function as community officers providing basic police services in their geographical jurisdiction. The approach is problem-oriented and almost always in conjunction with other (relevant) partners (Inspectie Politie, 1997).

At the time of this study, community policing had been either partially or fully implemented in 15 of the 25 regional police forces.[5] The form it has taken and the results achieved vary between forces, due to a lack of clarity as to objectives and goals (Inspectie Politie, 1997), and it is yet unclear which form of community policing should serve as a prototype for the rest of the regional forces. In essence, there is no one perfect form of community policing, although one can speak of general guidelines and assumptions. It is small-scale, community- and problem-oriented, and the task of an officer is individual responsibility for the provision of services in her jurisdiction. Tasks, responsibilities, powers, and procedures must be clearly delineated and agreed upon. Good communication is essential. A thorough exchange of information as well as networking with other agencies involved in the provision of security is critical (Inspectie Politie, 1997).

Philosophy

To fully understand and appreciate the development of crime prevention in The Netherlands, it is important to understand the philosophy guiding the policy. On a very concrete and operational level, the foundation of Dutch crime prevention policy is built on two concepts. The first is "safety consciousness"—the process of convincing "(t)he community, its institutions and their agents, ... of their own role in preventing crime." The second element is pragmatism (Sorgdrager, 1997). Because many of the causes of criminal behavior lie in spheres over which the police and courts have little or no control, it is necessary to mobilize numerous and various organizations in the fight against crime (Kohnstamm, 1997). Since 1985, hundreds of crime prevention projects have been planned and implemented and have involved the cooperative venture of numerous agencies, organizations, and groups. All projects are evaluated to examine their degree of effectiveness (Sorgdrager, 1997). On a more abstract level, however, a variety of concepts or theories can provide the basis for crime prevention and community policing projects and programs.

The Concept of "Broken Windows"

Based on an experiment by Philip Zimbardo in 1969, Wilson and Kelling (1982) coined the term "broken windows theory." In the experiment, Zimbardo abandoned cars in two distinctly different neighborhoods: one, a lower-class borough in the Bronx (New York City) and the other in a middle-class neighborhood in Palo Alto, California. Ten minutes after the car was abandoned in the Bronx, it was vandalized. Twenty-four hours later the car was totally stripped and randomly destroyed. In contrast, the car remained intact for a week in the relatively well kept neighborhood in Palo Alto. Zimbardo finally took a sledgehammer to the car, and within a few hours residents or passersby had totally destroyed the car. Wilson and Kelling (1982) refer to this as the "broken windows theory." One broken window starts the chain of events resulting in further vandalism and destruction. "At the community level, disorder and crime are usually inextricably linked, in a kind of developmental sequence" (Wilson and Kelling, 1982:281).

Wilson and Kelling (1982) argue that once neighborhoods are touched by disorder, nonserious offenses, followed by more serious crimes, are likely to occur. Therefore, in order to reduce or prevent crime, attention must be paid to fairly minor issues pertaining to disorder. Addressing problems surrounding disorder improves the quality of life in a neighborhood, thus increasing residents' perceptions of subjective safety.

Communitarianism

Communitarianism puts forth the notion that individuals are not autonomous and that they do not exist in isolation. They are shaped by the values and culture of communities. It argues for the revival of community norms and goals and for the community's understanding of its own responsibility. It calls for a reduction of the "welfare state" and for the individual, family, and agencies within the community to assume some of these tasks. The state (that is, justice agencies, police, utilization of the criminal law) should be the last to be called in to handle problems (CPN Tools, nd).

One way to improve residents' perceptions of safety in their neighborhoods is to create a more positive living environment. Communitarianism would argue that the work must begin with the people in the neighborhood. Whatever they cannot do for themselves must be assumed by social work agencies, housing associations, schools, the local government, and—when all else fails—the police and justice agencies. Community policing's emphasis on permanent geographical locations allows the police to become intimately familiar with the residents and problems of their neighborhood. Through intense contact with the residents, the police are able to establish social networks to assist in solving the problems that cause unsafe conditions (Vleesenbeck et al., 1996).

A movement that is gaining in popularity in The Netherlands over the last decade, and particularly within the past few years is that of "promoting self-reliant behavior." This goal was made popular by Dutch police psychologist Frans Denkers in the 1970s. It is defined as

> The ability and willingness of individuals, in conjunction with others, to find a solution to problems or conflicts, from the viewpoint that it is not the sole responsibility of the government—or the police—to solve problems between citizens, but also the responsibility of the citizens themselves (Gunther Moore and Van der Vijver, 1997).

The Society, Safety and Police Foundation has been working for a number of years to encourage and stimulate citizens to become more self-reliant regarding security issues. Research funded by the foundation and conducted by the Institute for Applied Social Science at the Catholic University of Nijmegen found that in order to stimulate citizens to take initiatives in coping with security issues, a number of criteria had to be met. The police had to be physically near, present or easily reached, available (that is, reacting promptly and taking citizens' complaints seriously), and reliable (even with minor problems). The police also had to know and be known to the residents. It is important for the residents to feel confident that if they are unable to handle a situation, the police will quickly and effectively intervene (Gunther Moore and Van der Vijver, 1997).

Social Control Theory

A theory complimentary to communitarianism, and one which was the basis for numerous delinquency prevention projects (and the idea behind the approach in the society and criminality policy), is that of social control theory (Terlouw, 1991). Control theory (Hirschi, 1969) argues that all young people have the potential to become delinquent and that some do not become involved in criminal behavior due to their attachment to conventional society (that is, through their law-abiding friends, parents, school, church, jobs). This theory argues, in essence, that people must be integrated into society, that their social bonds and commitment to societal institutions (family, friends, schools, jobs) must be strengthened, that moral values and respect for others must be instilled, and that involvement in conventional activities must be increased. It is not difficult to discern a convergence between communitarianism and social control theory, nor to understand how these theories influence crime prevention policy.

Application of Theory to Practice

Despite the fact that these theories are closely related, they serve as the theoretical underpinnings for different crime prevention projects. The concept of "Broken Windows"—the idea that nuisances (in Dutch, *overlast*) are often responsible for peoples' fear of crime and influence residents' perception of their security—is the theoretical basis for numerous neighborhood-based prevention projects (for example, to clean up neighborhoods, provide better lighting, and find solutions to problems caused by drug users and loitering youth). Communitarianism, and more importantly the movement toward inducing self-reliant behavior, also play an important role in the primary prevention projects aimed at specific neighborhoods, building complexes, or other locations. Community police officers are frequently involved in improving contact with and between residents and encouraging them to identify and participate in solving relevant problems. Together, these efforts form the basis for the majority of the community-oriented primary prevention projects aimed at improving the quality of life and reducing the fear of crime.

Although Social Control theory is tangentially related to Communitarianism, it serves as the theoretical basis for secondary or tertiary prevention projects aimed at eliminating delinquency—those projects which focus on either at-risk youth (those showing signs of problems but prior to their involvement in serious crime) or those who have already come into contact with the police or criminal justice agencies. These young people often lack good societal integration. They have poor or troubled relationships with parents, problems in school (high truancy and dropout rates), and weak bonds with other positive role models or social institutions such as the workplace or religious institutions. The majority of projects aimed at preventing delinquency focus upon strengthening the bonds of youth to society by providing them with better educational and job opportunities and improved relations with their parents.

Existing Crime Prevention Projects

Crime prevention, or the programs which are an outgrowth of this concept, can be divided into four categories. In the true sense of the word, it can involve "pure" or primary prevention. This is intervention prior to the commission of a criminal or delinquent act. Primary prevention is aimed at the public in general. Secondary prevention is aimed at high-risk groups. Potential success is based upon the ability to identify individuals at risk or those who are exposed to crime-inducing situations. Crime prevention also may involve intervention aimed at a specific individual after that person has committed a crime, targeting for instance the hard-core offenders to prevent them from recidivism (programs targeting those in prison). Tertiary prevention, in a sense, is a form of "post prevention" (Van Erpicum, 1997). The fourth category of crime prevention measures rely heavily upon technology-oriented prevention. Even though these measures may be initially successful, say, in reducing shoplifting in a particular store, they do not address the root of the problem and often result in displacement of the offense.

This section will deal with innovative projects and programs in The Netherlands. They will be divided into categories dealing with "pure" prevention projects (primary and secondary), methods of targeting youth, and ways of working with offenders already in the criminal justice system (at an attempt to prevent recidivism). For each major section, one or two projects will be described. This section closes with a discussion on prevention through technology.

Primary and Secondary Prevention Projects

Community Agent New Style (Utrecht)

"With two feet in the neighborhood" is the umbrella slogan for a new type of service provided by the police department in the Utrecht region. With three independent but integrated projects (community police, emergency assistance, and service), the Utrecht police hope to increase objective and subjective safety in the region. These three projects have been implemented in all districts within the city. By improving the efficiency and effectiveness of emergency assistance and service, it is possible to reduce the number of officers involved in these tasks and thereby increase the number of officers involved in community policing. With a new seven-step training program (Action Learning), and the philosophy of "learn by doing," neighborhood officers (and their supervisors) are trained to identify local security problems, figure out the most effective measures of addressing the problems, coordinate effectively with other agencies and organizations in the neighborhood also dealing with the issue of safety, and involve residents and motivate them to become self-reliant.

The training is provided by the National Center for Safety and the Quality of Life attached to the police training institute De Boskamp. Community officers, their supervisors, and other members of the police department, as well as service providers from other organizations involved in safety issues, participate in the training guided by the philosophy of "bottom up." The action-learning training was nominated in 1996 for the Police Innovation Award (Stuive and Belt, 1997).

The Drieviant Project

This project combines the strengths of the community police officer and the "improvement" worker in a socially deprived area in the city of Deventer. The goal of the project, initiated in January 1995, was to increase the residents' responsibility in dealing with their own problems. This is one of seven projects throughout the country in which pairs of police and social workers join forces to stimulate the residents to make improvements and address security issues in their neighborhoods. Initial contact is established with professionals in the area (doctors, teachers, pharmacists—those who work in the neighborhood and have sufficient contact with residents). Personal phone calls are made, and letters are sent out inviting participation in the "movement." The first task is to assess the problems. Then groups are formed (to be involved with traffic, residences, and activities), each with the task of solving problems related to a particular area. Despite the fact that it is difficult to keep motivation levels high throughout the project and that a number of earlier participants withdrew from the group, they still profited from the fact that they are now aware of whom they must approach to get something accomplished, whether it be the police, the social service agencies, the building association, or the municipal government. Relations between the residents and the aforementioned organizations has improved since the implementation of this project (which won the 1996 Police Innovation Award). It has also shown that a cooperative working relationship between two parties, which traditionally viewed each other with skepticism and mistrust, is not only possible but profitable (Boerstra, 1997).

Targeting Youth

The Ministry of Justice, in its attempt to reduce juvenile crime, has introduced a number of themes related to crime prevention among juveniles, including giving extra attention to offenders under the age of 12, dealing aggressively with hard-core groups of criminal youth, increasing the ability to enforce the education requirement (and reduce truancy), improving the operation of the "youth work guarantee plans," reducing violence on television, ensuring that crime prevention is high on the agenda of regional and municipal youth policy, and increasing the "youth expertise" among the police (Jeugdcriminaliteit, 1998). The Public Prosecution Department also focuses on youthful offenders. Besides reducing the time to process juvenile cases through the system, the department is attempting to increase the community service sentence by 10 percent, and to pinpoint high-risk groups (juvenile repetitive offenders, hard-core youth, first offenders, and at-risk youth under the age of 12), precarious locations, and the various types of offenses. Improving relationships with other justice agencies (such as the police, Child Protective Services, and Probation) is also a focal point of current policy (Versterking Jeugd Openbaar Ministerie, 1998).

A more intensive cooperative relationship between the police and Child Protective Services is expected within the coming years. In the past, most formal police reports were not passed on to the Child Protective Services (approximately 30 percent in the large cities). The goal was to increase this to 100 percent by the year 2000 in the hope of identifying potential problems and realizing an early intervention (*Intensiveren samenwerking Raad en politie*, 1998).

Youth Prevention Project South-East Brabant

This project, initiated by the South-East Brabant Police force, is aimed at intervening in the lives of youth at risk who have come into contact with the police, in an attempt to prevent further delinquent behavior. The pilot project began in 1992. The idea is that since the police are often the first to come into contact with the families, they should step in rather than wait for other organizations to intervene at a later stage. Police officers involved in this project receive a special 5-day training period and have clearly established job descriptions. An interactive support network and registration system has been set up. The project is a client-oriented, pragmatic, integrated approach involving schools, community workers, and mental health services. Individual plans of action (focusing on the family, school, and leisure activities) are developed for each client and are established with the client, the parents, police, and other agencies such as schools, community workers, and mental health services. The program, which lasts approximately three months, contains a "well-researched, step-by-step operational procedure" aimed at developing the child's skills so that he can cope on his own. The program has resulted in an 84 percent rate of nonoffending (far surpassing the 70 percent target rate). Participants (both clients and network partners) reported a high degree of satisfaction and significant improvement in their living situation. The project, one of the European Crime Prevention Award nominations, has received national attention and has been replicated in other cities (Ministerie van Justitie, 1997).

Job Training Fair

Because a large number of juvenile offenders (often from minority origins) come from deprived neighborhoods with few chances for a successful future (many are school dropouts with no job skills), the Amsterdam-Amstelland Police force initiated a program that would tackle the problem at its roots. To increase safety on the streets, ethnic youths were encouraged to finish their education in order to increase their chances of finding a job. The police approached a school in one of the working class areas that provides secondary education to more than 1,000 pupils, 99 percent of whom are from ethnic backgrounds. In cooperation with multinational corporations, many of which have excellent training programs or job placement opportunities for ethnic minorities, a job fair was organized. Moral and financial support was provided by the local employment office, the Chamber of Commerce, the council executive, and the foundation Working Together for a Safer Amsterdam. Successful members of the ethnic minority community gave motivational (positive criticism) speeches. Whether these young people actually obtained jobs as a result of the fair is not known, but it did serve as a motivational force and orientation point for them and has served to strengthen the bonds between youth, schools, police, private industry, and local government. This project was nominated for the European Crime Prevention Award 1997 (Ministerie van Justitie, 1997).

Tertiary Prevention Projects

Vocational Training Project in De Havenstraat Penitentiary

This project was begun with the goal of reintegrating ex-offenders and reducing the 70 percent recidivism rate, which is common among those having served time in a correctional facility. Begun in 1990, this program provides prisoners with a special four- to five-week vocational training course during the last stage of their detention. Ten inmates participate in each training class. Some course modules take place outside the penitentiary. These courses have been developed in consultation with industry so that inmates receive training in areas that are sought after by companies. Trainers often come from different companies, and during a particular part of the course, managers and personnel directors are invited to give presentations about their organization. The companies guarantee the inmates a job at the end of their sentence. Upon completion of the course an inmate may be released. Professional guidance is also provided for assistance in housing (an integral part of the project) and budgetary concerns. The results of the project have been gratifying. "For those taking part in projects where work, accommodation and social support were arranged, the level of non-reoffending was as high as 70%" (Ministerie van Justitie, 1997). This project was the winner of the 1998 Hein Roethof Award.

New Perspectives

In 1992, the municipal government of Amsterdam began working closely together with five city districts, the Public Prosecution Department, and the police and municipal Moroccan Council to address the problems of marginalized (predominantly Moroccan) youths[6] involved in criminal behavior. "New Perspectives" is a methodical approach involving individual, short-term intensive supervision. Its approach is practical, positive, and solution-oriented. The intensive intervention period lasts usually between two and four months. Intervention workers are on call 24 hours a day and manage a maximum caseload of four clients. Young people are assisted in finding jobs or are required to attend school. A network of various agencies works to support this project. The Moroccan community is also represented. After the intensive supervision period, the young people are transferred to the network of agencies or to a particular institution within the network. Attempts are made to restore ties to the family and community.

The project has seen some success. Approximately 72 percent of those who began (and 84 percent of those who completed) the intervention reported improvements in various areas. Perhaps most remarkable is that 78 percent are no longer involved in crime; 62 percent report improvements in education, 60 percent in family relationships, 57 percent in relations with friends, and 56 percent in leisure-time activities. Despite the fact that the program is not completely without problems, the success thus far has resulted in its expansion to other districts in the city (Veenbaas and Noorda, 1997). This project has become the spearhead of the policy for Youth and Security in the city of Amsterdam.

Technology-Oriented Prevention Projects

An Umbrella Organization: Senter

Under the auspices of the Ministry of Economic Affairs, the Senter organization was established to develop new technologies that would contribute to solving societal problems. At a workshop in January 1998, Senter introduced nine projects. A number dealt with improving security in and around the home via the television cable as an alarm system (experiments are being conducted in numerous cities but for now the price is quite high). The future focuses on the possibility of securing the home via the Internet. An experiment in Delft, Socially Safe Lighting, examines which type of lighting would invoke an increased sense of security among residents.

Two other projects focus on auto theft. In one, microchips were installed in motors and approximately 4,000 cars have been outfitted with these chips. Police officers, aided by a special apparatus, can read the chip (although different equipment is needed to read the chips from different production companies) and thus obtain information about the car's owner. Another project, the Location Determination System, resembles the British tracing and tracking system making it possible to locate stolen cars. As soon as the starter interrupter is damaged or destroyed, an automatic alarm signals in the call room. The location of the car is then determined, and the owner is contacted to verify the situation.

Another pilot project in The Hague is experimenting with the use of biometric identification (unique characteristics such as fingerprints) for the (future) Citizen Service Card. Another possible use would be to identify individuals at major events (such as football competitions) who were penalized with a stadium prohibition and thus prevent them from attending the game. Utrecht is experimenting with allowing victims of shoplifting to file an electronic complaint with the police. The last project is an attempt to establish an electronic database of the "best prevention practices" (Van Erpecum, 1998).

Camera Supervision in Buses

This project, which involved placing visible and hidden cameras in public transportation buses, was designed to reduce vandalism on buses, to enhance passengers' objective and subjective safety, and to increase the probability of arresting offenders. In conforming with privacy laws, passengers were notified when boarding that cameras had been installed in the bus. This project was begun in the transportation region of Voorne-Putten in 1995. Although cameras were installed in hundreds of buses, not all of them were actually recording. The results of the project show that offenses declined as a result of the video films. Vandalism in the buses was reduced, and passengers reported feeling safer and experienced a decrease in the number of undesirable encounters with other passengers. Bus drivers reported a decrease in aggression with a weapon, pickpocketing, vandalism, and theft. The cleanup team reported a decline in graffiti, and no displacement from the buses to the bus shelters was observed. All those involved reported such positive results that plans are being made to expand the project to Zeeland, South-Holland South and Ijsel-Lek (Ferwerda and Geutjes, 1997).

IMPLEMENTATION OF AND CHALLENGES TO CRIME PREVENTION PROJECTS

Although crime prevention projects vary widely in terms of design, intended target groups, and the means to achieve results, they all share one thing in common: They were born of necessity to address a particular problem. Certain qualities are characteristic of crime prevention projects in The Netherlands. Almost all of them are cooperative ventures involving more than one agency. Interestingly, organizations and individuals that previously held diametrically opposed views (such as police and probation officers or police and social workers) are now beginning to work closely together to tackle common problems. Funding for projects comes from numerous sources (besides from the central and municipal governments), as even private industry is realizing that prevention is cheaper than replacing stolen merchandise and cars. Projects undergo evaluations, and successful ventures are replicated in other cities. Various justice agencies previously associated with repression (the police, the Department of Public Prosecution, Probation, and Child Protective Services) are systematically introducing "prevention" into their policies.

Despite an abundance of projects and optimism concerning their success, there are a number of challenges. In addition to the most common problems of funding, of coordinating various agencies, and of keeping residents motivated to participate in long-term projects in their neighborhoods, there are other problems. Some of these problems are discussed below.

MEASURING THE EFFECTS OF LONG-TERM RESULTS

Although many projects boast success based on the number of individuals taking part or completing the project, little longitudinal research has been conducted on the long-term effects of such crime prevention projects. Additionally, recidivism rates are limited measures for gauging the success of a project. For example, continued addiction to drugs and receiving welfare payments while not perpetrating crime indicates only a partial reintegration into society. It is necessary to conduct qualitatively good evaluations. This means conducting both a process evaluation (to determine whether the project is being implemented as it was designed) and an effect evaluation (to determine whether the project is achieving expected goals). Longitudinal studies to measure long-term effectiveness should become a priority.

PROBLEMS SURROUNDING AN INTEGRATED APPROACH TO SAFETY

Integrated safety projects, widespread throughout The Netherlands, require the cooperative input from a number of different organizations. Research into the success of this method of working indicates that participants are highly motivated and enthusiastic about new-found partners (Bruinsma and Aronowitz, 1997). This approach, however, is not without problems. Kleiman and Terlouw (1997a) found that in some instances, integrated work constituted nothing more than ad hoc or limited, rather than structured, contact between partners. Another problem arises when partners rely on one another to accomplish goals. When one organization falls short on its responsibilities, the entire process is delayed. Better coordination and improved channels of communication can help alleviate this problem (Bruinsma and Aronowitz, 1997).

Cultural Differences Within the Target Group

Crime prevention projects that try to reduce the marginalization of foreign youths emphasize strengthening the bonds with the family and other societal institutions. Projects often must cope with language, religious, and cultural differences. These problems have been overcome by working closely with ethnic organizations and community leaders in both the planning and implementation stages (*Justitiekrant*, 1998).

Identifying Risk Factors and Individuals

Many secondary prevention projects aim at "reintegrating" marginalized youth, or youth at risk. In order to be successful, it is important to be able to identify risk factors in the child, family, and environment (Junger-Tas, 1997a and 1997b) that are most strongly identified with later criminal behavior. Unfortunately, no one factor is a predictor of future criminality, and only when several factors coexist is there an increased risk of serious health, welfare, and crime problems. Furthermore, predictions of human behavior are based on probabilities, not on individual cases.

Voluntary versus Involuntary Participation

Does a society have the right to force individuals to undergo "treatment" or alternative sanctions? This remains an ethical question, but research has shown that court-mandated participants in projects for "hard-core" youth are more likely to complete the program than those who are voluntary clients (Kleiman and Terlouw, 1997a). Junger-Tas (1997a) argues that forced intervention should not always be ruled out.

Issues of Privacy

This issue is particularly strong in the discussions surrounding the exchange of data and the use of cameras (that is, how the data and pictures are used and by whom). A balance must be struck between the necessity to protect property and maintain public order, and the individual's right to privacy. Regarding the use of cameras, the Ministers of Justice and of the Interior recommended that the public must be informed about the use of the cameras and that individuals must have a legitimate reason or need to utilize cameras. Restrictions are placed on the use of cameras by individuals to protect their property, and permits are required to direct cameras at public places (Offens, 1997).

THE FUTURE OF CRIME PREVENTION IN THE NETHERLANDS

Crime prevention in The Netherlands must focus upon a wide range of topics. Three important foci demand special attention. The first is preventing juvenile crime. The second is concentrating energies on making neighborhoods and cities safer. The third major category deals with serious forms of crime such as cross-border, international, and organized crime.

Preventing Juvenile Crime

Junger-Tas (1997a) identifies three conditions which must be present to prevent juvenile crime. First, identify risk factors in the individual, the family, and the environment, and strengthen protection factors, such as individual characteristics, social bonds, and social support (Junger-Tas, 1997a:102). Second, effective methods must be available to help children develop in a "prosocial" direction so that they are able to cope without turning at a later point in their lives to crime or drugs. Third, there must be a consistent and comprehensive prevention policy that produces clear middle and long term, as well as short term, results (Junger-Tas, 1997a and 1997b).

The reduction of risk factors through numerous projects and agencies has, to a degree, been accomplished. Despite abundant projects addressing juvenile delinquents and marginalized youth, Junger-Tas (1997a and 1997b) argues that no consistent integral policy (with the exception of Child Welfare Clinics, which offer care for mothers and children for the first two years after birth) and that projects work in isolation from one another maintaining their own register of clients and satisfying their own aims. What is missing is an interconnection between the different projects (Junger-Tas, 1997b:23).

Creating Safer Neighborhoods

The Netherlands continues to emphasize the importance of subjective security: The need for people to live free of a fear of crime and to enjoy a sense of safety in their neighborhoods and cities. Integral, community safety plans need to be rational and based upon problem analyses. Concrete, preferably measurable, goals must be established, projects must be implemented, and follow-up studies must be conducted to determine the success of the measures employed.

It is important to create a good infrastructure of organizations and services. Since individuals and neighborhoods at risk rarely manifest isolated problems, it is wise to involve the services of numerous and varied organizations, including schools, mental health services, social workers, employment offices, private industry, housing authorities, the police, and criminal justice agencies. Crime prevention teams comprised of members from various organizations at the local level are best able to tackle problems in their own communities. These teams should include members of minority groups living in the area. When and where possible, community residents should be motivated to identify and solve problems in their neighborhoods.

Preventing Cross-Border, International Organized Crime

However, crime prevention, even "pure" prevention projects at the local level, is not a panacea for preventing *all* types of crime, particularly national and international organized crime. Pure repression is not the solution to the problem either. Even with this type of serious crime, prevention measures can and must be taken to prevent organized crime from infiltrating the legitimate business world and corrupting members of business and government. Law enforcement agencies can work closely, for example, with the Chamber of Commerce and employment centers to check the background and reliability of candidates, and to deny the issuance of licenses and business or building permits. This preventive approach, however, still relies heavily upon contact with the local authorities and the involvement of numerous organizations. A combination of prevention and repression is the most effective way to deal with many problems. The challenge lies in achieving the proper balance.

International Cooperation

A structural exchange of information and cooperation at an international level is critical to battle large-scale, cross-border crime. Even when addressing the issues of less serious crime problems, there is a need to exchange practical information in the form of "best practices" in crime prevention. International conferences on crime prevention [such as the European Union Conference on Crime Prevention (Noordwikj), The Netherlands, May 11-14, 1997] and international crime prevention exchange programs (such as the one sponsored by the Dutch Ministry of Justice in The Netherlands from March 30 to April 2, 1998) (Van Oostveen, 1998) are vital, initial steps toward information sharing and a structural approach to battling crime in the countries.

SUMMARY

The internal European borders opened to trade with the European Union legislation in 1993. A single currency will be introduced in the year 2002. With open borders, fewer restrictions and limited internal controls at the frontiers, E.U. countries have witnessed changes in the types and seriousness of problems (an influx of both legal and illegal aliens) and criminal offenses (organized crime and the trade in humans). Despite differences, common concerns include providing a safe environment for citizens and making people feel secure in their neighborhoods and cities. Through an expanded exchange of information, it may be possible to aid one another in achieving that shared goal.

It must not be forgotten that the cities belong to the residents. They need the opportunities and structures to invest in their communities. Attention must be paid to subjective measures of safety (fear of crime) which can often be increased by addressing, at the local level, such minor problems as graffiti, garbage on the streets, or poor lighting in public areas. Small improvements can make big differences. It is important to achieve immediate results (improved lighting) as well as long-term success (the integration of marginalized youth). Only a wide palette of crime prevention projects, combined with a tempered degree of suppression, will achieve maximum results. Effective research into the success of these projects, in particular longitudinal research to determine the long-term results is critical to ensure that prevention is on the right path.

ENDNOTES

1. Municipalities (148) with a population of over 40,000 had a municipal police force; smaller municipalities were served by the national police.
2. Regional forces are divided into a number of divisions or districts, headed by a district commander; districts are subdivided into smaller units.
3. The section on *Grotestedenbeleid* is taken from Aronowitz (1998), with permission from the publisher.

4. *Availability* is defined as responding to an emergency call within 15 minutes and as being available according to arrangements. *Accessibility* is defined as being able to contact a police officer either in the station or by phone (*Inspectie Politie*, 1997).
5. This figure, however, is based upon the definition used by the Police Inspectorate. Haarlem, for example, utilizes the team policing approach. Since officers are not permanently assigned to one specific area, the Inspectorate does not consider this to meet the criteria of community policing. Another smaller jurisdiction, on the other hand, has one police officer for approximately 30,000 residents. Because of the permanent assignment and responsibility for this area, this does meet the criteria for community policing.
6. Ninety-two percent of the youth are males between the ages of 14 and 20. Almost two-thirds are of Moroccan origin. Because the program could have a positive impact upon younger delinquent youth, the minimum age has been lowered to 10 in two city districts.

Chapter 6

Some Thoughts on the Relationship Between Crime Prevention and Policing in Contemporary Australia[1]

Rick Sarre

Since the formation of specialized police forces in Australia (see Finnane, 1994, for further information), a key philosophical question has persisted in much of the policy debate surrounding the functions and powers of the police: whether it is preferable for police to operate in a primarily *reactive* fashion, that is, to assist when called upon to do so and not otherwise, or whether the police should act in a *proactive* fashion and seek to prevent crime before it happens. The former model of operation is based in classical theory, which is suspicious of intervention into the lives of people who do not warrant state attention. Moreover, it accommodates the fact that disorder often occurs on private property which means that the police have to be called in rather than assume an open invitation. The latter model of operation reflects the old adage that an ounce of prevention is worth a pound of cure. It sits comfortably with the positivists' belief that it is possible to seek out preferred targets for special policing and thereby thwart crime that otherwise might have been committed. Although proactive policing often excites civil libertarian concerns, advocates point to its potential for cost effectiveness and deterrence

Clearly, modern policing in Australia is based upon a mixture of the above two models. Indeed, some commentators have pointed to the spurious nature of the *reactive-proactive* distinction, "given that police make a host of prior decisions about organizational structure and practices, patterns of patrolling, and dispositional options which influence the capacity of police to observe or respond to particular incidents of trouble well in advance" (James and Polk, 1989:47). Nevertheless, it is possible to discern an historical growth in reactive emphases in Australian policing in the latter half of the twentieth century because efforts were made to improve police efficiency, patrol response times, and "clear-up" rates. Planners have sought to employ innovative technological advancements to aid the crime-fighting task.

Yet there has been little discernible arrest in rising rates of crime. Even highly "professional" police find that increased knowledge about specialized policing methods, quicker response times, and reliance upon crime control expertise appear not to reap the rewards (lower crime rates and less fear of crime) they were touted to bring.

Other commentators point to the fact that crime is, for the most part, outside of the control of the police (see, for example, Robinson, 1989:173). Patrol and investigative practices can have no real effect, directly or indirectly, on some of the root causes of offending, such as poor urban conditions, political divisions, dysfunctional families, poorly socialized individuals, adverse corrections policies, ill-conceived laws, and failed economies. The legal and physical powers of police can have little effect upon the sociological, educational, and political milieu in which we live (Homel, 1994). It is simply not possible for police to forestall all the precursors of aggressive and dishonest conduct and social unrest. Similarly, police effectiveness can be only as good as the general community's willingness to accept it (O'Malley, 1979:272).

At the same time, and notwithstanding the disparaging picture painted above, Australian governments have substantially increased resourcing of police activities in the last two decades. As of June 1995, there were 42,517 sworn officers in all of Australia, and 9,820 support staff (Mukherjee and Graycar, 1997:66). The current cost of public law enforcement in Australia is now well over A$3 billion annually, a figure which does not include federal police nor the Australian Protective Services. Spending on policing has increased substantially in Australia since 1960. Compared to other claims upon scarce public finance reserves, policing now, arguably, demands a disproportionate share of government revenues. Faced with increasing ineffectiveness and higher costs, police and public policy makers have been aware of a looming crisis for some time.

A number of approaches have been made by policy makers to address the concerns over ineffective policing and inappropriate costs. These approaches can be described as having manifested themselves in two key ways over the past twenty years.

First, many governments adopted a strategy of policing that allowed for other forms of specialist social "ordering," such as the role played by private police forces and security providers, fee-for-service security personnel, ad hoc protection services and in-house security forces (Sarre, 1998; Prenzler and Sarre, 1998). As one commentator has noted, the modes and methods of "private" policing of yesteryear can be now said to have survived the formation of public forces (Johnston, 1992:6). The move toward privatization has been both revolutionary and quiet. More recently, private security personnel have been embraced more positively as potential (albeit junior and more cost-effective) partners in the collaborative production of community safety. Private policing is a crucial component of policing now and in the future, but it is a phenomenon outside the immediate focus of the discussion that follows. Second, governments have adopted a notion of policing that rejects the traditional crime-fighting role of police in favor of a new approach which embraces "crime prevention." This response provides the focus of the discussion that follows.

This chapter describes the various manifestations of crime prevention as they have been embraced by police and then explores the reasons why the policy is, if not essentially flawed, at least compromised by the difficulties inherent in its conceptualization, implementation, and evaluation. Readers ought not be surprised that one conclusion drawn at the end of this chapter is that it is possible, if not distinctly advantageous, for some aspects of the crime prevention task to proceed without police involvement at all.

CRIME PREVENTION IN AUSTRALIA

It is difficult to define adequately the notion of "crime prevention," although the definition of Van Dijk and De Waard provides a good start. They refer to "the total of all private initiatives and state policies other than enforcement of the criminal law, aimed at the reduction of damage caused by acts defined as criminal by the state" (1991:483). This definition has appeal because it excludes essentially reactive responses to law breaking, and includes proactive crime prevention systems and partnerships that may be embraced by police from time to time. In other words, it is not possible to argue, on this definition, that *everything* that the police do is "crime prevention," a line often touted by some police policy makers. By way of example, there are two broad initiatives and policing policies that could fall within the Van Dijk and De Waard definition. The first is a more specific strategy developed from a North American model and often referred to as *problem-oriented policing*. The second describes a more general strategy, widely implemented currently in Australia, known as *community policing*. Following the descriptions hereunder is a discussion of their place in the crime prevention picture generally.

Problem-Oriented Policing

Problem-oriented policing (sometimes referred to as "problem-solving" policing) is a phenomenon first developed in the 1970s in the United States. Rather than passively wait for distress calls or patrolling at random, under this model the police anticipate calls and identify and classify local crime and disorder problems with a view to preventing their repetition. There has been a great deal of academic and police interest in this model of policing in Australia (West, 1992:195; Beyer, 1993:136; Budz, 1998). Under the traditional "reactive" model, police respond to calls which are not only treated in isolation but considered "closed" when the case is solved or shelved. Problem-oriented policing, by contrast, searches for precipitating factors, which, if eliminated, may stop or at least limit the antisocial conduct that otherwise might have occurred. Such an approach is consistent with the view that not all crime and disorder problems are the same, nor are the neighborhoods and communities in which they occur the same. If, for example, the police determine that a majority of calls for assistance come from a small minority of addresses, then it would make sense to seek out the root causes of the trouble to address or to eliminate it. As one study of crime prevention (in South Australia) concluded,

> Research has shown that many calls for police assistance are related—for example they stem from the same locations, families or groups of offenders—

> but officers attending the scene often do not realize this. Police need to be retrained in techniques of analysis and be given the resources, flexibility and authority to confront underlying causes (Sutton and Fisher, 1989:28).

A problem-oriented strategy, therefore, attempts to collate incidents in order to paint a larger picture by thinking about the nature of the problem rather than just responding to it. A community with a history of violence may benefit from a policing strategy that endeavors to locate the causes of the violence (for example, if there is an ongoing feud between rivals). Policing strategies may be tailored to suit a community largely comprising indigenous people or enclaves of refugees. The determining feature need not be geographic. A police domestic violence strategy may attempt to tailor the police response according to information on the family and in addition the potential for violence through access to firearms by the potentially violent member of the household.

The evidence is anecdotal, but one Australian study (Beyer, 1991) of the four years of the Frankston Police Community Involvement Program was cautiously optimistic of success, but only if changes were made to the management, supervision, promotion, and training structures within Victoria Police.

> For problem-oriented policing to succeed, current administration and leadership would need to be changed. For instance it would need to be value-based rather than focused on detailed instructions. Police instructions have been designed to prescribe police action in every eventuality and their effect has been to push "underground" the initiative of police and the encouragement of the unproductive police philosophy of "just stay out of trouble". The new management approach would assume workers care about the substance of their work, and would recognize the importance of informal leadership, of resourcefulness and peer influence. (Beyer, 1991:98).

Beyer's conclusion was that, given these changes, and given a far more participatory role for all officers engaged in the task, the results can be "valuable" (1991:99; also 1993:26ff).

Community Policing

Theorists and practitioners have long believed that it is simply impossible for the police to carry alone the tasks formerly carried out by peacekeepers of yesteryear. There has been a deliberate endeavor therefore to broaden the role of the community in providing its own security. The term often used is "community policing." Defining precisely what is meant by the concept is difficult (Sarre, 1996). Nevertheless, it can be said with some confidence that community policing places key emphases on foot patrols ("beat" policing) and on community service performance not unlike the roles played by community self-help vigilantes of the preindustrial period. Furthermore, and perhaps most importantly, community policing shifts the philosophy of policing away from reactive measures toward proactive models of operation, most often after a process of community consultation.

In many instances the police and the community embrace this concept together, in theory at least. Crime prevention programs developed in consultation with the community seek to develop a "partnership" during the life of that particular program. These programs can be large-scale, for example, Victoria's Integrated Anti-Crime Strategy, launched in 1991 with its "consultative committee" concept (Beyer, 1993:15) and South Australia's Together against Crime Strategy, designed to facilitate implementation of crime prevention at the local level in consultation with the police (South Australia, 1990; Sutton, 1991 and 1994; Van Dijk, 1990). Queensland's youth-based community initiatives; the Western Australian Crime Prevention program, a school-based program in the Northern Territory; the antidrug campaign Operation NOAH (an acronym for narcotics, opiates, amphetamines, and hashish); Operation Paradox, which targets sexual abuse of children and operates currently in at least two states; the Safety House program; and the Confident Living programs for the elderly are all examples of statewide community policing schemes and liaisons in operation currently in Australia (Byrne, 1992; McMillan, 1992; Nixon, 1992; West, 1992).

Crime prevention in consultation with the community can also operate at the microlevel. Specialist squads for groups in crisis or for groups regarded as marginalized have been set up in some jurisdictions in Australia (Grabosky, 1992:258). In addition, the very popular USA-inspired Neighborhood Watch program (first tried in Australia in 1984) now extends to Rural Watch, River Watch, Business Watch, School Watch, Mobile Watch, Marine Watch, Pub Watch, Industrial Watch, Witness Watch, Hospital Watch, and even Commuter Watch.

Police have also been keen to provide assistance to concerned groups with advice on how to create conditions less conducive to opportunistic crime. Known as "situational crime prevention," these initiatives involve police advising on "opportunity reduction" and "target hardening" techniques, general topics outside the immediate focus of this chapter but discussed extensively elsewhere in the Australian criminological literature (Geason and Wilson, 1990).

However, as the section that follows describes, what these cooperative endeavors offer in theory may not hold up in reality. A number of theoretical, philosophical, and practical difficulties confront police and governmental policy makers as they proceed in their quest for a greater emphasis on crime prevention.

POTENTIAL DIFFICULTIES FOR POLICE CRIME PREVENTION MODELS

In the rhetoric that usually accompanies crime prevention programs and initiatives such as those described above, the police are most often described as the facilitators and motivators of these programs in consultation with other members of the community. The police, in this view, are but mere participants in crime prevention initiatives that are shared with members of wider society. There is a concern, however, that in reality the police find it difficult to see their role as crime preventers. Rather, the police prefer to be seen as crime fighters. There is a general unwillingness to share their duties with the community. Very little evidence exists to counter the suspicion that crime prevention endeavors are predominantly police-driven, operate with a minimum of consultation, and are put into practice for purposes largely unrelated to "the reduction of damage caused by acts defined as criminal by the state." Police wish to remain pivotal in all aspects of their functions. The following section describes how these misgivings have arisen. The difficulties that emerge are then explored and counterarguments, where appropriate, are presented.

The Crime Prevention Role is Misunderstood by Police

What is suspected by many observers is that the police see their crime prevention role as little more than a public relations exercise designed, more than anything else, to enhance their public image and as a tool for engendering popular support. Consciously or subconsciously, police assume that any program that renders the public better disposed toward the police must, per se, be good crime prevention and they assess its success and effectiveness accordingly (Sutton, 1994:224).

On the definition of crime prevention presented above, however, the phenomenon involves more than merely currying favor with the public. It involves incorporating into police life programs and strategies, other than the enforcement of the law, that will bring about a reduction in crime. Although there is nothing sinister in police seeking a better public profile, it is the view of some commentators that the lip service paid to crime prevention has less to do with reducing criminal damage and criminality generally and more to do with providing a political milieu in which a return to a more conservative political agenda— "neoliberal" reforms (O'Malley, 1994:286)—is likely. Such an agenda would extend to giving police powers to intrude into areas they could not formerly enter or to attempt to follow the USA-styled so-called zero-tolerance (Dixon, 1998). It would be an unusual Australian police force indeed that was not constantly seeking enhanced powers of investigation, arrest, questioning of suspects, and fire power.[2] There is perhaps a perception among the police that these powers are unlikely to be enhanced, and "tolerance" reduced, in an environment where crime prevention is made the chief focus of their work.

Police Find it Difficult to Consult

The police, essentially, decide what aspects of the crime prevention task ought to be shared, if any. They have a preferred access to the media, and theirs is the view that dominates. They alone determine in a top-down fashion which sections of the community are qualified to accept responsibilities. Listening to what the community has to say is not the same as consultation.

The problem begins with police structures, which operate through a chain of command. Such an authority structure is, by its nature, exclusive and makes community participation very difficult, because it is not possible to put a member of the community into that command structure. In relation to Neighborhood Watch, for example, many police officers embrace the concept as a useful vehicle for public relations,

> as long as it is a largely one-way flow of information *to* the police. As soon as demands are put *on* the police for changes in priorities, it may be seen as an intrusion (e.g., Etter, 1995). There are very real limits to which the police, as presently organized, are capable of responding to local concerns (Brown, 1987: 252).

Furthermore,

> [w]hile public cooperation is sought, active citizen intervention is discouraged. . . . Not only does the definition of the problem remain with the police, but information about it comes through their perspective (Wilson, 1986:69).

Analyses of the British experience with consultative programs also paints a bleak picture. Police continue to set the agenda and use the forum they have created to explain and obtain support for the decisions which they have made (Stratta, 1992).

One's suspicions would be confirmed in the jurisdiction to which the following police sergeant was responsible. His explanation of the way the Neighborhood Watch process works is as follows.

> Community consultation in this context is not groups of civilians trying to direct police activities, it is simply developing a forum for civilians to improve their communications with police (Coster, 1988:62).

The experience of police-community liaison in Victoria from 1991 to 1992 was monitored by the Federation of Community Legal Centres. In their

May 1993 report, they noted that local committees never really possessed decision-making power. It was always subject to directions higher in the force (Biondo and Palmer, 1993:8). The police pretend to share policing functions in order to appease those who wish to be consulted yet also to locate a scapegoat in the event of failure. The fact that an earlier "consultation" by the police took place provides an excuse for the breakdown (O'Connor, 1988:55).

By the same token, it should be noted that a cooperative model of policing may not always be appropriate. One argument maintains that although citizen involvement in the policing function is important, there would be little disquiet if that involvement were circumscribed on some occasions, for example, where justice and humanity are at risk of citizen compromise (Grabosky, 1992:266). Furthermore, another argument maintains that the police have the best resources for putting the entire picture together, because localization of the problem tends to limit the focus of participants to the extent that they miss the "big" picture. However, on balance, and since approximately 80 percent of policing functions are not of the crime-fighting type (Beyer, 1993:126), it would appear that this counterargument has many limitations.

If Consultation Occurs, Police Consult Only Select Groups

There is a strong suspicion that police consultation on crime prevention issues takes place only with those groups in the community who have an interest in the outcome. The consultation process is, indeed, most attractive to those who wish to protect their vested interests in the prevailing social order. Certainly the rich, the better educated, and the occupationally dominant would tend, one would think, to be most involved. That being the case, the information flowing to the police is skewed in favor of those who stand to gain from police involvement, and the programs reliant upon that information are unlikely to be the most appropriate for the situation at hand.

> In a homogeneous society, mutual penetration of police and citizenry may be in the public interest. But in a heterogeneous or otherwise divided society, those interests which are generally dominant may also be expected to dominate citizen participation in law enforcement. Unequal participation need not lead inevitably to unequal justice, but the potential shortcomings of participatory imbalance are such that certain checks or compensatory devices may be required. . . . Private citizens are not always motivated by a sense of public service or civic virtue (Grabosky, 1992:266).

Those who have no voice have little incentive to consult with police. There must be serious concerns about any consultative process that provides benefits only in one direction and against those who are frequently at the harsh receiving end of police authority, where there is less likelihood of consultation and more likelihood of a sullen acceptance of power.

> The community has for a long time been used to the notion that the police carry the major responsibility for dealing with the consequences of many of society's ills. Although governments, the police, and a number of community organizations are beginning to proclaim that crime prevention is fundamentally a community responsibility, we still have a long way to go before that is generally accepted (Smith, 1993:8).

Indeed, where is the "community" of which the police often speak? Social groups and networks are not geographical any more. There is no guarantee that a particular community will have a single voice, assuming that a community can be identified. It is far more likely that they will have a dozen or more voices. Even though there is much evidence of community consultation with Neighborhood Watch block captains, there is much less evidence of consultations by police with young unemployed people, with ethnic and racial minorities, with women, with the elderly, and with children at risk. Police relations with indigenous Australians have been at a low ebb for many years. Since the findings of the Royal Commission into Aboriginal Deaths in Custody were published in 1991 (Royal Commission, 1991), some efforts have been made to remedy the situation. But there is some distance to go.

> The truth is that most police just do not know how to communicate with Aboriginal people. . . . If the managers at the top level are not aware of the specific issues that they are involved in, then it becomes extremely frustrating for those people who are on the ground dealing with problems every day to understand and accept uninformed decisions of senior managers (Crawford, 1992:8).

Moreover, it is foolish, if not dangerous, for the police to hold out consultation as a means by which they can promise universal approval. The police may be faced with the problem of trying to fulfill completely contradictory expectations. What does a police force do when different groups require different means to address crime prevention ends? One would suspect that they capitulate by being more concerned with placating key interest groups (for example, local business interests) than with confronting them in order to seek long-term solutions. Where is the evidence, for example, that police see their role as confronting traders with their concerns about the possible criminogenic effects of their practices? One might suspect that police would prefer to curtail the activities of drunken patrons as they spill onto the street, rather than attempt to stifle the late-night trading practices of the local hotel.

It should not be forgotten, however, that in no small part there are many ways in which community needs concerning housing, education, health, and security (especially for women and children) have been recognized after a process of negotiation with police. There is some evidence in Australia that many groups display great admiration for the ability of police to assist them in determining outcomes that are favorable to themselves (Veno and Veno, 1989:145). Although these groups are not just those made up of the "dominant" or "privileged," this will be the rule rather than the exception.

The Police Would Never Allow the Sharing of "Their" Work

How much sharing of policing tasks is possible? For Sutton, the answer involves another set of questions:

> To continue to be effective in crime prevention, police will need not just to review program philosophies and contents but to learn to work in inter-agency and community-based partnerships which may involve relinquishing much of the direct control now taken for granted (Sutton, 1994:226).

Although these sentiments are laudable, they are unlikely to be carried out in practice because the policing organization has a distinctly legalistic view of authority. To have the police share some of their roles with the community would be akin to asking them to redefine their entire raison d'être. It is therefore highly unlikely that there would be a genuine shift to a community partnership for crime prevention without a great deal of reluctance and stalling, if not at the senior management level then certainly at the level of the rank-and-file patrol officer. It is even less likely that police management would allow their budgetary management to shift to the community.

What little research has been done on so-called interagency cooperation indicates that dominant players are unwilling to step to one side and assume a secondary role in determining implementation needs.

> One of our most consistent findings is the tendency for inter-agency conflicts and tensions to re-appear, in spite of co-operative efforts, reflecting the opposition between state agencies at a deep structural level. We have also found consistent and persistent struggles between local authority departments over limited resources, power and prestige (Sampson et al., 1988:482).

This interagency difficulty intensifies when powerful groups like the police see their territory being threatened.

> The police are often enthusiastic proponents of the multi-agency approach, but they tend to prefer to set the agendas and to dominate forum meetings,

> and then to ignore the multi-agency framework when it suits their own needs (Sampson et al., 1988:491).

This confirms the fear of theorists that crime prevention, or indeed any other police task, is unlikely to be successfully shared, especially if the view that the security of society is directly related to the allocation of resources to police is persistently presented by police and politicians in a "law and order" political climate.

The Results of Cooperative Endeavors are Equivocal

Evaluations are notoriously difficult to conceive, to carry out (Cunneen, 1991; Sarre, 1991a, 1991b, 1992, and 1994), and to bring within reasonable budgets (Sherman, 1992). Even assuming that it is possible to isolate satisfactory measurement criteria (Beyer, 1993:126), the evaluations that have been conducted do not strongly recommend many crime prevention projects. The conclusion that many researchers are reaching is that neither "more of the same" nor marginally different tactics are likely to improve police crime prevention effectiveness to a substantial degree (Reiner, 1992:155).

These findings present a marked dilemma for policy makers and commentators alike. For if the picture is bleak, then those police who favor no shared responsibilities with the community confirm their argument. In an alternative view, the evaluations that have been done have either failed to concentrate upon appropriate measurement criteria or have merely highlighted the problems in implementation. These problems are often exacerbated by the reluctance of the police to be involved or of the community to be aware of their potential involvement. In other words, evaluations must now target the potential of police consultative crime prevention efforts to be successfully implemented where there is an incentive for them to work. In other words, poor evaluations do not necessarily, to date, indicate that the theory is flawed.

Police Culture Makes it Difficult if not Impossible

The police culture is one that is difficult to change and one that "needs to be acknowledged in all its complexity and, at times, perversity" (Goldsmith, 1990:92). Many police reject the notion that theirs is to be a service organization in which the public are seen as clients who consume police services. Status and promotion within the police ranks are still strongly linked to successes in crime fighting, not in some amorphous concept of crime prevention. The police culture continues to undervalue skill in prevention, excellence in community relations, and victim support (Beyer,

1991:95). So, although lip service is often paid to the crime prevention role of police, it is unlikely to be embraced fully by the rank-and-file officer. Moreover, the changes required to alter this description of police culture are unlikely to come about while there are few rewards within the structure of the force for those who embrace crime prevention eagerly. What is needed is a complete overhaul of performance criteria and the granting of promotional rewards for those who embrace a community consultative model (Chan, 1997). The difficulty for determining criteria for rewards, awards, and promotion, however, is that crime prevention is not as easily tangible and as readily assessable as is disarming an armed offender or saving the lives of victims of a siege.

SUMMARY

A new emphasis on the crime prevention role of police by Australian policy makers during the decade of the nineties came as no surprise to observers. For a long time it was known that, compared to other public services, the police in Australia acquire a disproportionately large share of public finance. In addition, the evidence of truly effective policing is patchy indeed, if crime rates are anything to go by. Faced with this conundrum, policy makers have sought out new methods of policing that emphasize crime prevention and the requirement of consultation between the police and the public they serve. To a large extent the "problem-oriented" and "community policing" foci of the nineties were indicative of this trend. But the evidence suggests that what may be sound in theory does not always emerge in practice. Suspicion is mounting that there is little consultation between police and the community in relation to crime prevention, and if the consultation does take place, the coordination is dominated by police. Evaluations paint a bleak picture. One may suspect that the enhanced lip service being paid to the notion is merely a part of a broader political emphasis upon appeasing the public and ushering in a society in which police maintain their status as the sole guarantors of order.

Far preferable for the future would be an agenda that shares crime prevention responsibilities between public and private bodies and between specialists and nonspecialists—professionals and laypersons alike. Indeed there may be some crime prevention functions that the police should avoid altogether. In many instances the preferred option might be for the police to simply ensure that other agencies and groups (such as health care agencies, schools, town planners, and families) take responsibility. Shared policing options should be implemented and evaluated as a matter of urgency in Australian jurisdictions. The mix of private and public, specialized and nonspecialized, police and community involvement in crime prevention will change over time and in accordance with a variety of factors. This change cannot occur while police continue to see the crime prevention task as theirs alone or as a mere public relations exercise. One suggestion is that governments take a lead role, implementing a policy of appointing a Minister of Policing rather than a Minister of Police.

The ability of all order-maintaining functionaries to share their given responsibilities and to see the wisdom in recognizing their limitations will determine to a large extent whether communities will be able to effectively handle — within reasonable budgets—the challenges posed in the years to come.

ENDNOTES

1. Reproduced in part from Sarre (1997) "Crime Prevention and Police" in P. O'Malley and A. Sutton (eds.) *Crime Prevention in Australia: Issues in Policy and Research.* Annandale: Federation Press. Used with kind permission of the Federation Press.
2. It should be noted that the Australian police have extensive powers to demand name and address, to ask people to move on, and to enforce "good order" offenses such as consorting, offensive language, and offensive behavior (Dixon, 1997: Chap. 2).

SECTION 2

This second set of chapters deals with countries that are relatively young or are in transition from colonial status to independence. Each of these countries has been greatly influenced by western forms of policing and social control, particularly because of their past connections with large colonial powers, most prominently the British empire. Each of these countries was under the control of an outside force for an extended period of time. As such, their formal policing and law enforcement methods reflect the actions of their earlier colonial masters.

At the same time, these countries and the people who dwell in these areas did not exist in a vacuum prior to their colonial status. Various forms of social control were in use, many of which relied on the individual, family, or neighbors to provide protection and response to injurious behavior. Although not always formally codified, the traditions and folk customs of the people outlined the appropriate responses to deviant behavior. These traditional methods did not disappear when colonial powers took control, but persisted in the smaller, more isolated communities as the only form of social control. The efforts of the newer, formalized policing systems were restricted mostly to the larger towns and to the business interests of the ruling colonists. The demise of colonial rule has seen movements to accommodate both the traditional forms of social control and the newer police enforcement methods.

These chapters discuss recent community policing and crime prevention efforts as they relate to the history of the individual countries. Throughout the materials a great deal of discussion focuses on the challenges of maintaining social control in countries whose citizens are ethnically diverse. There is a clear effort to use that diversity to enhance social control.

CHAPTER 7

CRIME PREVENTION: THE COMMUNITY POLICING APPROACH IN ISRAEL

Ruth Geva

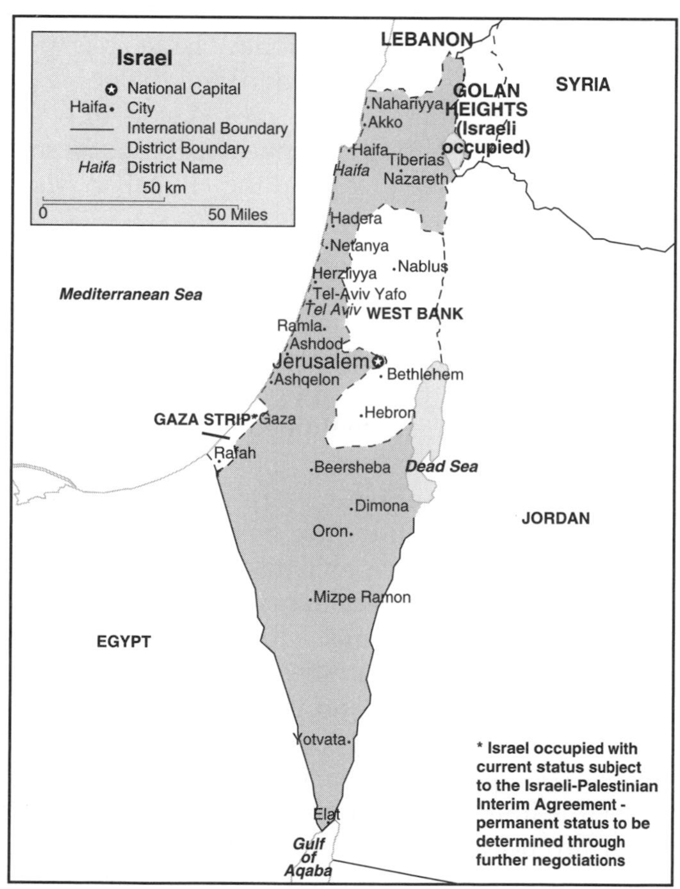

Although the state of Israel is a young country with a short history of law enforcement and a criminal justice system that is only fifty years old, rapid development has, unfortunately, created problems that are similar to those encountered in more established societies. This chapter focuses on efforts being made by the Israeli Police Service in the area of crime prevention and control, but attention is also directed toward the specific use of community policing as a central strategy. It reports, in brief, on far-ranging activities in social welfare and on the mobilization of the public for voluntary action in crime prevention. The chapter also touches upon the international activity being made to further crime prevention in Israel.

The Social Setting of Israel

Israel is a relatively small country of approximately 27,800 square kilometers (7,992 square miles). All of Israel's immediate neighbors are Arab countries, predominantly Muslim. Today, Israel has peaceful relations with its two neighbors, Egypt and Jordan, and the peace process is slowly advancing with the Palestinian authorities. A state of war still exists with Syria and Lebanon, and fighting continues in the southern part of Lebanon between the Israel Defense Forces and the Hizballah Muslim fundamentalist groups.

Constant terrorist incidents, initiated and carried out by extremist Palestinian terrorist groups, occur within the borders of Israel. Since its inception, Israeli society has had to cope with the tensions generated by the intermingling of diverse population groups, the struggle for economic independence, the constant threat and outbreak of war, terrorist activities, and concern for the country's continued existence and the security of its population.

Internally, friction exists not only among Israel's various ethnic groups, but also between ultraorthodox religious Jews and secular ones. This tension occasionally causes civil disorder and rioting requiring police intervention. Since the country's independence in 1948, Israel has absorbed waves of mass immigration from the four corners of the globe: Holocaust survivors, entire Jewish communities from the Islamic countries of the Middle East and North Africa, and large waves of immigration from the former Soviet Union and Ethiopia. Unemployment and the problems of cultural and social adjustment have produced outbursts of anger and frustration among the immigrants and problems for the youth. The weakening of patriarchal authority, especially within the traditional communities, has caused serious adjustment problems, especially among young people—often expressed in criminal activity.

CRIME TRENDS

The crime rate in Israel held relatively steady for the past fifteen years, but the last three years have shown a steady climb, especially in property crime and vehicle theft. Violent crime, including domestic violence, has also been on the rise during the last few years. In 1997, the total reported crime rates were 6,220 per 100,000 population (some 370,000 crimes reported). A steady increase in the population (presently almost 6 million)—due to natural growth, as well as to immigration—was not accompanied by a complementary increase in police services. The police now find it difficult to deal with rising calls for services. Other major changes that have occurred in the last five years include the strengthening and infiltration of organized groups of criminals (many from abroad) who are using Israel as a base for money laundering, gambling, prostitution, and drug trafficking.

A particularly problematic area of criminal activity in Israel involves property crime, which constitutes about 75 percent of all reported crime to the police. Car theft has been on the rise for the last five years and has proved especially difficult to combat. Burglary and thefts of all kinds constitute a serious problem, both in terms of economic impact on society and the traumatic effect these crimes have on the victims. Because of the anonymous nature of these offenses, very few criminals are actually caught. A certain feeling of helplessness is experienced, both by the law enforcement community as well as by the average citizen, whose fear of crime also has risen.

The increase in serious crimes, especially violent incidents such as armed robbery, has caused great worry among the criminal justice community. An increase in reported cases of domestic violence, youth violence in the schools, and in drug use by young people has the criminal justice system looking for innovative approaches to reduce crime and increase the quality of life for its citizens.

THE ISRAEL POLICE

A national service, the Israel Police receives all its financing from the national budget. There are no other police forces (local or otherwise) in Israel. The mission of the police is to "prevent crime and to provide safety and security to the public it serves, working to improve the quality of life of the community." The Police Service includes over 25,000 police officers spread out around the country. The headquarters of the police is in Jerusalem; six districts, 12 subdistricts and 72 stations are responsible for policing services around the country. A typical police station employs roughly 100 to

200 police officers in patrol, traffic, investigative, intelligence, and Civil Guard (volunteer) sectors. No civilian staff exist, at this point, although the multiyear Strategic Plan of the police is to civilianize posts gradually, thus providing more positions for professional policing functions. The Civil Guard volunteer police provide preventive patrol and support to the police station, particularly within the local neighborhoods. Over 400 Civil Guard bases allow some 40,000 volunteers to receive training and equipment for their work as volunteer police officers from a "close-to-home" center.

The Police Service is also responsible for all border patrol and control functions, riot control, and antiterrorist activity that takes place within the boundaries of the country. Most of these functions are provided by the paramilitary Border Guard Police. A special antiterrorist team, as well as a national negotiation team, belong to the Police Service. Other national units include the Fraud Unit, the Serious Crime Unit, and the International Crime Unit. Since the Israel Police force is responsible for antiterrorist activity and for security of the public, considerable effort and resources are engendered to prevent terrorist action and to provide an increased sense of personal security at the neighborhood level. Between 1987 and 1994 (including the years of the *Intifada*, or Palestinian uprising, during which hundreds of terrorist incidents per year occurred) the Israel Police greatly increased the number of units used to maintain public order. These units included mounted police to keep order in Jerusalem, Border Guard police platoons to assist the army in the border areas and in the Administered Territories, riot control units, motorcycle units, and bomb disposal personnel.

THE PREVENTIVE APPROACH WITHIN THE ISRAEL POLICE

Until the late 1970s, the police force in Israel was typically a reactive service providing mainly investigative and patrol functions. Furthermore, the police were assigned, through legislation, to handle internal security (that is, antiterrorist activity) within the borders of the country, thus deflecting the focus of the police function from a crime orientation to a security orientation. The police, however, have developed and maintained several elements that are the cornerstones to an integrated crime prevention approach. These use the community-based policing strategy as a base.

CORNERSTONES TO AN INTEGRATED PREVENTIVE COMMUNITY APPROACH

Preventive Patrol as Deterrent

In the early 1970s, due to terrorist actions, a grassroots organization demanded that organized patrols be set up to patrol the neighborhoods. The founders of this organization believed that these patrols would provide both preventive activity as well as a means for emergency call up of security forces. This was the beginning of the Civil Guard voluntary police, which has since been integrated with and managed by the police (Gal, 1993).

Preventive patrol, along with searching bags and cars, was and still is considered one of the main preventive strategies relating to deterring terrorist activities. Since most of the terrorist acts until the start of the *Intifada* occurred when the terrorists placed bombs in public places, patrolling neighborhoods, searching buses and bus stops, and having security personnel check the bags of visitors to stores and public recreational areas is considered the most effective way of deterring the potential terrorist. The Civil Guard volunteers, who police the neighborhood on foot or on car patrol, were originally recruited (in 1974) specifically for this purpose. However, over the years, it was realized that this massive resource (ranging from 40,000 to 50,000 volunteers) could be utilized for preventing other criminal activity. Today, the volunteers are used in all policing activities, including patrolling, setting up roadblocks, surveillance, publicity dissemination, searching for lost persons, and for order maintenance in public gatherings of all kinds (Gal, 1993; Geva, 1995c).

It is difficult to ascertain what preventive effect these volunteers have on crime. However, almost every day, offenders are caught by volunteers on patrol around the country. In various small communities, where citizens have rallied and patrol units exist, the number of evening and nighttime offenses has dropped dramatically (CPU, 1997).

A State-of-the-Art Survey on Crime Prevention

In 1976, a parliamentary commission (The Shimron Commission) was set up to look at the rise in crime and to suggest ways in which the police and other governmental organizations can work to prevent crime. There was a feeling that much fear of crime (sometimes unfounded) existed. This commission suggested, inter alia, that the police should offer the public information about how citizens could better protect themselves against crime and provide data on victimization and crime patterns. Citizens could then be more aware of which crimes might affect them and could act accordingly to reduce the possibility of becoming a victim. In 1978, the Israel Police decided to undertake a major survey (Geva, 1981) focusing on how crime prevention is tackled by other police forces around the world and to suggest ways the police could better work with the public in this respect.

The Jerusalem Pilot Project in a Residential Community

Based on the survey, a pilot project in one Jerusalem neighborhood was implemented during the period 1980–1981 (Geva and Israel, 1982). The project targeted property crime, especially burglaries. The community Civil Guard volunteers were mobilized to assist the police in disseminating crime prevention information and marking property. Lectures on crime prevention techniques were given to citizens, an exhibition of antitheft devices was set up in a supermarket, and home visits were made to give security advice and to undertake surveys of premises. The pilot project was accompanied by an evaluation comparing the results to a similar control neighborhood that was not given any such treatment. The success of the strategy appeared in a crime rate drop of approximately 35 percent. About 38 percent of the public changed their behavior in the direction suggested by the police, and those receiving advice felt more secure.

New Preventive Functions

Following the Jerusalem pilot project, the Police Commissioner decided to implement four functional instruments within the police service: neighborhood police officers, crime prevention officers, a Community Relations Unit, and the Crime Prevention and Security Unit.

Neighborhood Police Officers

These officers would give advice to the public, provide policing services within neighborhoods having no police station, and work with the community on crime prevention projects. These neighborhood officers have diminished in number greatly over the years but, in places where such officers still serve, they have been reoriented into Community Policing officers. An evaluation of the effectiveness of these neighborhood police officers showed that there was an increase in the public's sense of security in those areas where they worked (Friedmann, 1986).

Crime Prevention Officers

It was also decided to train experts on crime prevention techniques (mainly target hardening) within the police stations who would provide citizens with advice on how to take better care of their property. These officers were not exclusively crime prevention officers, although they received specialized training and functioned as such while performing other police functions. By 1984, these functions were passed on to the Civil Guard police officers, who, together with volunteers, gave advice, organized property marking operations, and gave lectures to the public (especially to the elderly and other at-risk groups) (Geva, 1992).

The Community-Police Relations Unit

This unit was set up in 1980. Its job was to find ways to increase the good relations between the public and the police. Using "Police-Community Days," the functions of the police were displayed, publicity material was disseminated, and media relations and various activities were targeted to increase public cooperation with the police, mainly through volunteering to the Civil Guard. Pamphlets and other crime prevention material were also published by this unit, although the contents were decided upon by the CPSU (described in the next section). In 1983, this unit became part of the Civil Guard department and remained there until 1994 when these functions became integrated into the newly founded independent Community Policing Unit.

The Crime Prevention and Security Unit (CPSU)

Set up at headquarters, this unit's job was to analyze crime patterns (mainly in the area of property crime) and to find ways, together with various public and private organizations, to increase "situational crime prevention." The unit worked with the Israel Standard Institute to standardize equipment such as locks, alarms, and doors. It also worked with the National Insurance Association to decide on security equipment requirements for insured home owners and car owners, with the Locksmith Association to improve locks, and with the Private Security Companies Association to accredit companies according to the type of alarms and control rooms they used and their working methods. The unit also acted as the headquarters' authority on licensing matters of businesses and on security equipment and techniques regarding "essential installations" and nationally important institutions. This unit still exists as the Security Unit within the Operations Department at police headquarters and emphasizes the situational crime prevention aspects of businesses, installations, and facilities. Almost no activity is targeted on private dwellings or property.

Since 1969, Israel has had a law allowing for the security regulation of high-risk businesses. These businesses must install security equipment in order to receive or renew their business license. According to the Law of Commercial Licenses (1968), the police are responsible for granting commercial licenses to those firms "which require supervision in order to prevent any danger to the welfare of the public and to render it safe from burglary and robbery." Today, dozens of different types of businesses and thousands of separate business establishments fall within this category. The effectiveness of this strategy appears in the decrease in gasoline station holdups, bank holdups, and offenses at diamond polishing factories (Mor, 1990; Geva, 1992). This "situational crime prevention" strategy was the main preventive strategy for property crime and for certain forms of personal crimes (sexual attacks of women) until the early 1990s, when the Community Policing partnership approach began.

The Beit Dagan Project

One of the turning points in implementation of the integrated community policing approach can, in retrospect, be pinpointed to a project in a small drug-infested community called Beit Dagan, in the mid-1980s. This community was suffering from heavy drug trafficking and a sense of helplessness and fear among citizens. The community, together with the police, Civil Guard volunteers, community workers, and treatment organizations, worked to evict the drug traffickers from the city by closing off the main roadways, searching cars, and working with the addicts to send them to treatment centers. Publicity brought the citizens, the police, and the municipality together. Very soon, the area was "clean," and there was a new feeling of regeneration and control among the citizens (Nir, 1989). A new police commissioner resolved to change the policing strategy to community policing and to have the police cooperate to a much greater extent with the municipalities, while looking for integrated approaches to preventing crime.

The National Crime Reduction Council (NCRC): Working Groups

In 1988, the Ministry of Police set up the National Crime Reduction Council, whose task was to bring about a coordinated approach to crime prevention through work with all relevant ministries and organizations. During the years 1988–1992, various "working groups" were set up to target particular crimes. The task of these groups was to analyze a particular crime, bring together all the relevant parties (both public or private), and work out a strategy that could reduce the incidence of the crime. The recommendations of these committees were supposed to be disseminated to the local level for implementation. However, this never took place (Geva, 1990a).

Two such committees were set up, one on vehicle crime and one on crime against the elderly (Shapira, 1990). Most of the recommendations at the national level were implemented, including changes in law to increase the severity of punishment for vehicle theft, for the scrapping of cars, and for changing the identity of a car. These were the direct result of the work of the committee (Geva, 1990b).

Unfortunately, due to other pressing matters under the responsibility of the ministry (especially the changes in policing due to the Oslo Accords—the agreements with the Palestinians), this ministerial function was put on the "back burner." For several years the council functioned only in a partial manner. The main activity of this council became the initiation and financing of various local programs for the security of the elderly and to coordinate crime prevention (mainly relating to drugs and violence) in schools.

In recent years, the NCRC has put an emphasis on research into national crime trends, on gathering and disseminating information on crime prevention activity taking place around the country, and on mobilizing local crime prevention activity. Special emphasis is placed on at-risk youths (Shimshi, 1997). During 1998, the council has decided to return to its "target hardening" activity dealing with property crime.

The Beer-Sheva Domestic Violence Project

This project, begun in 1989, emphasized the referral of offenders to treatment, using police to "force" offenders to go to a treatment center. The process involves the police, the social services of the city, and a voluntary organization (Naamat) that sets up the treatment services. An evaluation of the treatment processes showed a clear reduction in recidivism in spouse abuse (Geva, 1995d; Shalev and Yehezkeli, 1995). This multidisciplinary partnership model for crime prevention is the cornerstone of the preventive strategy used in the community policing philosophy (Gimshi, 1995 and 1997).

STRATEGIC DECISION: IMPLEMENTING COMMUNITY POLICING

In the mid-1990s, the former commissioner of police decided to initiate a fundamental change in the concept of policing, from a basically reactive form to that of community orientation. This conceptual change did not prove to be easy, since the police force has, over the last twenty years, slowly evolved into a security-oriented force in which antiterrorist and emergency matters related to ensuring public safety have become of paramount importance. Classical crime prevention has taken second place (Heffetz, 1995). With limited resources, the local stations have suffered from a severe lack of personnel and equipment. The local stations are constantly called to provide backup support to Jerusalem (because of the many demonstrations and VIP visits) and to conduct security checks on roads and highways in the surrounding areas.

Other than contact with victims of crime, delinquent youths (Aharoni, 1991), and some situational crime prevention activities, integrated crime prevention activity has become a rarity. Police officers in the stations are hard-pressed to provide services for the needs of the population. Investigative activity is minimal and mainly restricted to egregious cases and to those for which some evidence is available. Thus, the quality of police services has dropped over the years. Conversely, other public services in Israel have undergone a "revolution" and have worked to provide increasingly better-quality services to the public, including decentralization of national services to the neighborhoods. Furthermore, the cities and communities within them have become much more empowered. In many cities, community councils have been set up that act as "mini-municipalities," using the local resources and volunteers to decide upon and provide a wide variety of services. Furthermore, the local media have flourished and are used extensively to offer information and mobilize the community to take action.

It was felt that the police had to find a different style of policing that would allow them to provide better service to the public, to work with the public on crime prevention, to utilize the resources within the community that could assist in providing improved services, and to consult with the public on what areas ought to be tackled and in what order of priority. The community policing paradigm was considered as a means to address all these needs.

The main components of community policing, as conceptualized in Israel, include

- Controlling crime and providing community safety through partnerships with organizations and institutions in the community
- Working proactively to mobilize community, multiagency, integrated activities for the reduction of crime and fear of crime
- Emphasizing proactive crime prevention activity in the areas of publicity, education, and training
- Conducting problem analyses of crimes and quality-of-life problems that the community finds important and most relevant
- Insisting upon "real-time" law enforcement functions of the police, especially regarding street crimes that decrease the quality of life in the neighborhood
- Providing quality police service
- Involving the community in the policing activities through the ongoing interaction with the public (CPU, 1996a, 1996b)

Implementation at City Level

In order to bring about this change, the Community Policing Unit (CPU) was established to work with the various police stations and geographical headquarters (Gimshi, 1997). It was decided to start at the station level where the effects would be greatest and could be accomplished in the shortest time. Only after two years of local activity did the CPU change its strategy and start working at the headquarters management level, in order to integrate the top brass into the supervision and training of officers at the station level. Although this process has caused some resentment and lack of coordination, the entire force is now working toward integrating this new policing strategy.

Between 1995 and 1998, and under the direction of the CPU staff, the police and its primary partners within the municipal and community leadership took part in a citywide two-day Planning Workshop. During this workshop, the Mission Statement of the police station was conceptualized; the local problems and needs of the community were scanned, prioritized, and analyzed; and an initial strategic plan was set out. Public opinion surveys in the city were conducted prior to this workshop in order to sample the problems that needed to be tackled.

Individual steering committees were set up, using the problem-oriented approach, to analyze the problems and put into effect an operational multiagency model to deal with each problem. The problems targeted ranged from vehicle theft, domestic violence, and drug abuse to youth vandalism and noise disturbances in public places (Ben-Harush, 1997; Farjun, 1997; CPU, 1996b, 1997).

Development Processes within the Police Station

Training and development at the local level is supervised by one of the CPU's organizational development officers, who is linked to the station and city and assisted by a civilian management advisor. Using discussions, workshops, and steering-group sessions, they help the police officers and their civilian partners, internalize the changes in policing philosophy and organizational culture, and develop leadership roles within the station and the community. The training emphasizes the partnership approach, problem-oriented policing techniques, quality services, and measuring goal achievement. All activity is carried out in cooperation with relevant parties from the community.

Today, approximately fifty cities around the country have undergone the change to community policing. Already, many communities are going into their second and third year of implementation and have now begun a second stage of goal setting. The community decides which plans should be continued, which should be retuned, and which new problems ought to be tackled.

POLICING BY OBJECTIVES: INTEGRATING QUALITY MANAGEMENT INTO THE COMMUNITY-BASED POLICING APPROACH

Of late, the Commissioner of Police has decided that, according to the suggestions of the Police Strategic Planning Team, stations will plan their semi-annual and yearly activities based on the targeting of specific objectives. Each station, using the community policing methods of consultation with representatives of the public, identifies the crime problems and incivilities that trouble the public the most. These "problems" are translated into "objectives" (CPU and Strategic Planning Unit, 1997). Car thefts, vandalism, burglaries of business premises, domestic violence, drug abuse by young people, traffic congestion, and so forth are some examples of such problems. Usually, these problems arise in "hot spots" (where there is a large percentage of the crime), thus making the targeted objective less diffuse. The measures of success are also decided upon, emphasizing mostly crime reduction, although qualitative measures are chosen, such as increased satisfaction with police services. In order to reach these objectives, the station police officers work in partnership with the municipality, organizations, and leaders in the community to work out an interdisciplinary strategic plan—a "community policing model." The use of resources from all these organizations is one of the principles that maximize the potential of all partners.

All such strategies usually include the following components:

- Target hardening and "situational crime prevention" tactics (often using innovative technology)
- Education
- Publicity
- Enforcement and/or control (sometimes not necessarily police, but rather using informal control measures, such as parents and peers)
- Referral to rehabilitation or treatment

THE COMMUNITY POLICING STATION AND THE COMMUNITY POLICING CENTER

The local police station (except for some national services provided by the Israel Police Force) provides all the services needed for the citizen in her community. The Community Policing Station ought to provide those services in accordance with the needs and priorities of the local citizens. Of late, "ministations," called Community Policing Centers (CPCs), have been set up in order to provide better localized service to the public and to decentralize services. To date there are over 70 such CPCs working around the country. Specialized training is provided to CPC officers in problem-solving strategies by the unit. These centers, mostly manned by one police officer, work in tandem with the local community center or council, the neighborhood Civil Guard base, and the volunteers working in the area. Besides providing regular policing services to the public, the police officer also targets local problems and works to reduce them.

Models for partnerships with the community have already been implemented in various stations and CPCs in an attempt to reduce the incidence of crime and to get to the root causes of the offenses. The implementation of long-term partnership models makes it necessary to rewrite procedures and work out the divisions of responsibility. Still in the infancy stage, procedures are not always set immediately, but rather become integrated over time. Other activities, usually being led by the Community Policing Station together with the municipality and various community organizations, include setting up of victim assistance units, conflict resolution units, volunteer assistance units for the aged, the use of the local media to assist in fighting crime, school watch units and a "safe-school model" (Ben-Harush, 1997; Bruchman, 1997).

Other Elements in Implementing Change within the Police Service

Obviously, wide-ranging changes must be wrought within the Israel Police Service if the move to the community policing philosophy is to be lasting and permanent. It is not enough just to make changes at the police station level or within the management levels. Changes must become part of the organizational culture and processes.

Such modifications should include the initial selection of the police officer, training at all stages of an officer's career, and the criteria for evaluating police officers. All these factors should be adapted to the new method of policing, emphasizing the officers' abilities to communicate with the public, to solve problems, to look at related events and see the "big picture," to mobilize resources within the community, and to be innovative. The CPU works with headquarters to integrate these goals into the basic policing philosophy. Already, the community policing philosophy is part and parcel of all training courses. The participants in the Senior Management Level Course, as their major course exercise, must examine one of the stations undergoing community policing and analyze the various processes taking place.

Another major organizational change is planned for the next two years. The Strategic Plan for the Police Service has decided that a "flattening" of the organizational pyramid is essential in order to empower the local levels and provide more efficient and effective services to the public. Thus, some of the subdistrict headquarters will be eliminated, with the result that there will be only one supervisory level between the station and the national headquarters. This change will assist in speeding up the process of community policing implementation at the local level, while, hopefully, also providing more resources to the local station.

Two headquarters committees have been set up to work on Guidelines of Operations in Community Policing and on ethics. These committees have been initiated based on continued requests from the stations for guidance in new situations to which they have not previously been exposed, in lieu of changes in styles of operations. For instance, before implementation of community policing, minimal interaction occurred between the police, the municipality, and local organizations—both public and private. The separate operations of police and other service providers were quite clear-cut. Now that partnerships have formed, including the joint use of resources to implement various multiagency models, ethical questions have been raised among the police officers, such as Should and can the police receive resources from the municipality or some organization, even if it provides better services to the public? and How can sponsorship of joint projects be accepted, especially when it is provided by a private company, without it appearing as if the police are sponsoring the company?

These and other questions have to be resolved through the provision of guidelines and operational procedures so that police officers do not get involved in unethical activity. The police cannot fall back on existing procedures that are no longer relevant to the present working methods of the stations.

EVALUATION OF INPUTS, PROCESSES AND RESULTS

The CPU's Research and Evaluation Section provides the tools needed for evaluating inputs, processes, outputs, and impacts of the new policing activities. The evaluation of community-oriented policing has centered on measuring the factors listed below.

Inputs

- Resources for mobilizing the public
- Resources for activating the preventive strategy

Processes

These are mainly used by the CPU to see if the strategy used to make the change is effective, including changes in

- Levels of partnerships
- Police and public attitudes regarding their separate functions and responsibilities
- Planning processes

Outputs

- Increased feeling of public satisfaction for various services
- A real (reported and unreported) reduction in targeted crime and reduction in recidivism
- Greater sense of public safety and a reduced level of fear of crime
- Reduced public complaints against police officers

The above, as well as other changes in attitudes, are measured through a variety of questionnaires and through qualitative and quantitative measures. The CPU is attempting to be attentive to the needs of both the police officers and the community, and not overburden them with complex data collection. Procedures must allow for ongoing evaluation by the field units themselves, once they have been fine-tuned.

A major evaluation under way headed by Menahem Amir and David Weisburd from the Institute of Criminology at the Hebrew University in Jerusalem, funded by the Ministry of Public Security's chief scientist. This study is looking at the process taking place over a 3-year period. The final report was published at the end of 1999.

Initial Results

The CPU is in the process of establishing baselines from which it will be possible to evaluate changes over time. In some cases, comparisons have been made between "community policing stations" and "noncommunity policing stations."

The initial results seem promising. Resource sharing has already brought about increased crime prevention activity. Referral of offenders to treatment, for example, for domestic violence and drug abuse (Shalev and Yehezkeli, 1995), within the community has become commonplace, while coordination between the police and the community services has become more efficient. It has been found that a real decrease in recidivism has been achieved.

Although it is too early to arrive at definitive conclusions, there has already been a significant decrease in complaints against police officers throughout the force and specifically in the community policing model stations. The "attitude questionnaire" has also shown significant positive feedback regarding the target stations compared to noncommunity policing stations (Regev and Shoham, 1998). Levels of satisfaction in community policing service centers have been higher than in noncommunity policing centers and an increase in such satisfaction over time is already being felt (Geva, 1998). The results of the Hebrew University evaluation, however, show that the field police officers are not as involved with the community policing strategy as the senior officers in the stations and are not yet feeling the improvements (Weisburd et al., 1998). This year, special efforts will be placed on evaluating the strategies that have been put into place to reduce crime. Base-level measures were gathered in the late 1990s.

LOOKING TO THE FUTURE

A variety of areas need work in the short and intermediate time range. First of all, an increased use of "crime prevention through environmental design" (CPTED) is needed. This strategy has been introduced to various local urban planning committees, and it is hoped that the CPTED key guidelines will be adopted as conditional to authorization of design plans for public buildings, urban planning, and housing units. Second, increased use of technology in urban crime prevention is required. For instance, although the use of CCTV and other high-tech equipment is the norm in business premises, this equipment is not used extensively in city centers and other high-risk areas. A pilot project is being considered in Netanya (a northern city) and, hopefully, this will lead to other successful programs.

A third need is increased coordination between the public and private sectors in planning and implementing crime prevention strategy. The Israel Police has yet to decide on its policy regarding this kind of cooperation. Ethical issues have yet to be worked out on how the police can eliminate the possibility of partiality toward particular private enterprises while at the same time utilizing the resources of this sector to achieve preventive goals. Guidelines are now being established that will allow for this cooperation.

The increased use of the media in crime prevention efforts is another concern. Until now, the media has been only partially involved in these activities. The abundance of local media (like newspapers and TV programs) make it possible to reach the public and raise awareness of specific problems; encourage participation in crime prevention activity; increase cooperation with the police; and allow for communication between the police, other service organizations, and the public. Crimestoppers is just one of the programs that the CPU is trying to initiate on both the local and national levels.

Finally, training in crime prevention strategies and problem-oriented policing is required. There is a great need to train both the police officers and their civilian partners in tackling crime using "problem-oriented" methods (identification, analysis, strategy construction, and implementation and evaluation). These are not inborn skills; they must be learned and practiced. It is hoped that resources will be made available to make this training much more available within the community.

National Organizations Working toward Crime Prevention

It has become clear to criminal justice professionals that conventional methods of dealing with crime are no longer sufficient. There is a need for an integrated approach to dealing with such complex issues. The crime prevention approach, which combines legislation, education, social welfare activities, enforcement, target hardening, and the use of high-tech approaches, must be increasingly implemented in a coordinated manner.

In order to develop this integrated approach to crime prevention, various national coordination agencies (both governmental and nongovernmental) have been set up to work toward prevention through primary, secondary, and tertiary efforts (Geva, 1992 and 1995a).

The National Anti-Drug Authority is responsible for the development, promulgation, and coordination of strategies, both on the national and local levels, for the minimization of drug abuse. These efforts include providing information to the public at large and to youth, assisting in treatment and rehabilitation programs provided by the Ministries of Health and the Ministry of Social Welfare, mobilizing the Community Action Project for prevention activity within the community, working on research and information gathering from around the world, enhancing the enforcement aspects of drug-abuse prevention (for instance, providing sophisticated technology as well as dog training), and furthering legislation connected to drug abuse (The National Anti-Drug Authority, 1995a and 1995b).

The Prisoner Rehabilitation Authority is a national body that works to find ways of reintegrating released inmates into society, using the voluntary, public, and private sectors as partners. They provide the released inmate with vocational training and frameworks, such as hostels and halfway houses, to gradually integrate offenders back into society (Hoffman, 1995; Israel Prison Service, 1995).

The National Council for the Child works to advance the rights and welfare of children in Israel. Special emphasis is placed on the care and treatment of juvenile victims and legislation to guarantee children's rights (Kadman, 1995).

The Ministry of Labor and Social Welfare is responsible for the correction and social welfare of delinquents and other groups who are at high risk of becoming delinquent (Hovav, 1988 and 1995). The ministry provides services aimed at reintegrating young and adult offenders into society through the work of the ministry's Adult and Juvenile Probation Services. Of special interest is the Youth Protection Authority, which administers various closed and open institutions, hostels, and halfway houses within the community. These facilities emphasize the education and rehabilitation of youths who have been removed from their homes because of extreme distress (like abuse and neglect). The Youth Rehabilitation Service provides services for adolescents who have dropped out of the educational system. The chief instruments used to guide the youth away from delinquency are the rehabilitative, therapeutic community day centers, which emphasize work and study groups as well as vocational training. Various drug abuse intervention programs, coordinating the Ministry of Health's treatment and maintenance programs, provide a comprehensive community treatment model for drug abuse victims, ranging from diagnosis to therapeutic groups.

The Ministry of Education's Youth Advancement local units work to reintegrate alienated youth into normal educational activity, and place special effort on absorbing and integrating immigrant youths. These units use various intervention approaches, both group and individual, such as supplementary education, vocational training, seminars, training courses, and advice (Lahav, 1995; Levy, 1991).

SUMMARY

In many ways, Israel is on the right road toward integrating the community policing crime prevention strategy into activities sponsored by national and local organizations. It is a long, hard uphill climb, especially because Israel is changing the mind-set of the police organization as well as that of the public—neither of which is used to taking responsibility for crime or its prevention. The partnership approach is not a smooth one. It takes much effort and resources for initial mobilization, and even more to keep the partnerships alive and operating. The police are still experimenting with the Policing by Objectives management method in the community-based policing context. It is not yet clear whether this management tool will act as an impetus toward working with the community. Increased reporting rates due to cooperation with the public often lead to greater fear and concern. Those working in the community crime prevention field must be patient and not overeager in pointing out success. Achievement will come if the police continue on this path. In-depth planning, careful implementation of true partnerships, and learning from evaluation will no doubt bring about positive crime prevention results.

CHAPTER 8

A COMMUNITY POLICING APPROACH TO CRIME PREVENTION: THE CASE OF INDIA

R.K. Raghavan and A. Shiva Sankar

A country with a land mass one-third of the United States, India has a population of 950 million, three times that of the States. It has three major religions, and its people speak as many as 18 languages, not including several dialects that are peculiar to certain parts of the country. Nearly 48 percent of the population are illiterate, a factor that compounds the problems of governance, especially policing.

After wresting independence from the British in 1947, the country adopted a Westminster model of government. The head of state is the President who acts on the advice of the Prime Minister and the Council of Ministers. The country is divided into 25 states, which share power with the central government in a variety of spheres.

Under the Constitution of India, "police" and "public order" are the bailiwick of the state. That is, these are areas in which a state legislature alone is competent to make laws. Nevertheless, the federal government has established several federal police forces to meet special problems. These forces include the Border Security Force, Central Industrial Security Force, and the National Security Guard. The Central Reserve Police Force is a unified comprehensive organization under the command of the federal government, which assists state governments in times of large-scale public disorder.

Each state police force is headed by a Director General (DGP), assisted by several additional DGPs and Inspectors General. Every state is divided into a number of districts, each headed by a Collector. All other functionaries, including the Police Chief (Superintendent of Police), report to this Collector.

The basic unit of policing is the police station headed by a Station House Officer (SHO) who has the rank of Inspector or Sub-Inspector of Police. There are 137 police officers per 100,000 population. A medium-sized state, such as Tamil Nadu, has more than 1,000 police stations; it has a force strength of more than 80,000.

CRIME IN INDIA

With an overall crime rate of 654.3 incidents per 100,000 population in 1995, India figures among the many high-crime nations (National Crime Records Bureau, 1997). There is great demand on the police for measures to enhance community safety. Table 8-1 gives an overview of crime in India.

The most disconcerting feature of crime in India is the rise in crime over the past decade. Except for marginal drops in 1987 and 1993, the trend shows a rising curve. The number of violent offenses attendant with violence was 50,000 in 1953, but 245,000 in 1995. This represents a nearly 500 percent increase over four decades. This rise demands attention from police officers and criminal justice policy makers. What should be of particular concern is the fact that homicides alone increased 44.3 percent over

TABLE 8-1 CRIME IN INDIA

	1994	1995	Change, %
Total crime (in millions)	5.5	6.0	8.9
Crime rate (per 100,000 population)	612.5	654.3	6.9
Individual offenses (in thousands)			
Murder	38.5	37.5	-2.9
Rape	13.2	13.8	4.1
Robbery	23.9	22.4	-6.2
Burglary	121.5	116.5	-4.1
Theft	303.6	294.3	-3.0

a 10-year period. The corresponding figure for the United States was 4.8 percent.

It is appropriate to question whether police activities in India offer evidence that a preventive strategy exists. More germane to this chapter is whether the police have shown faith in community policing as a tool in preventing crime. But first, we must examine the literature that highlights features of community policing, especially in the United States, which are believed to support crime prevention.

COMMUNITY POLICING AND CRIME PREVENTION

Perhaps the most convincing of all arguments in favor of community policing is that this strategy enlarges the resources available to the police in their fight against antisocial elements. As Bayley (1994) states,

> The police cannot solve society's crime problems alone. They need the assistance of the public in warning off would-be criminals, in notifying them of crimes and potential crimes, and in providing information that leads to the arrest and punishment of criminals.

This is of great significance, especially in the developing countries of Asia, because one sees a constant widening of the gap between popular expectations and the manpower and equipment available to the police to cope with such expectations. The gap becomes more pronounced if one considers the growing variety and complexity of public order problems and crime, coupled with the reluctance of governments to pump more and more money into policing rather than economic development. As a result, police leaders have become convinced that rather than implore the authorities to give additional staff and other resources, it will be a safer investment to concentrate on engendering the physical support of the community, which is the direct consumer of police services. This strategy places its faith in *synergism* ("the simultaneous actions of separate entities which together have greater total effect than the sum of their individual efforts") that is so prominent in team sports and in the music world (Miller and Hess, 1994).

Perhaps a more valid argument that generally favors community policing, which particularly upholds its utility in the area of crime and its control, is that such a strategy enhances the legitimacy of police operational decisions. There is here a sharing of power that can heighten the community's trust in the police (Trojanowicz and Bucqueroux, 1990). Such trust comes about because, during its implementation, the police display a certain self-effacement that is refreshingly different from their normal reluctance to assume a subordinate role vis-à-vis the community. This posture becomes meaningful, especially in the context of crime, because it is the community, more than the police, that understands the problems peculiar to the neighborhood which facilitate the designs of the offender. In sum, this is an admission by the police that the community is well-informed and therefore better-equipped to evolve an anticrime strategy, enhance self-esteem, and promote the community policing philosophy.

A question that flows logically from the above scenario is, Why emphasize crime prevention and not crime detection? Do the police not need community support in solving crimes that have already occurred? Undoubtedly, the police in every country seek public cooperation during the investigation of crime. Appeals for care in preserving a scene of crime, to provide background information on a suspect, and to look out for fugitives from justice with a view to apprehending them have become so standardized that they are taken for granted. Within limits, such cooperation has been forthcoming. What, unfortunately, has not been evident is an awareness that the community has an equally important role to play in prevention. There is either a state of resignation that crime cannot be prevented or a genuine belief that efforts in that direction are solely a police responsibility. It is this state of affairs that explains police anxiety to educate the community in order to make them understand that crime is preventable and that they have a definite role to play in formulating and implementing programs which tackle factors that are conducive to crime. Programs addressed to crime prevention, therefore, are prominent for their bias toward educating the community at large.

Some Past and Present Crime Prevention Practices

Although "community policing" is an expression in use only for the past few years, certain practices associated with it, especially in the area of crime prevention, have been in vogue for a longer period. Some of these are projects associated with street lighting, property marking, security surveys, citizen patrols, crime reporting, and Neighborhood Watch programs (Miller and Hess, 1994). Each of these is addressed briefly below.

The traditional view is that adequate *lighting of streets* deters crime. Environmental design has played a significant part in buttressing this tactic. Although some research does not support the crime reduction benefit claimed, there is reason to believe that it at least helps reduce fear of crime among law-abiding citizens and enhance their sense of security (Miller and Hess, 1994). Also significant is the position of Lurigio and Rosenbaum (1986) that street lighting will be ineffective unless supplemented by police patrols and active citizen reporting of what they see.

Property identification projects encourage citizens to engrave identification marks and record a unique number on their valuable items of property. Additionally, they are given stickers to display on windows and doors warning possible burglars that valuables have been distinctly marked and that such information is on police record. A moderately deterrent effect is claimed on behalf of this tactic (Miller and Hess, 1994).

Crime prevention security surveys are closely linked to the strategy of environmental design. Here, trained police officers undertake a thorough inspection of offices and residential buildings and look for weaknesses in doors, windows, locks, lighting, and other factors which facilitate burglaries. In the view of Rosenbaum and Lurigio (1997), such surveys offer good results only when a system exists to follow up the police recommendations at their implementation stage.

The focus of *citizen patrol projects* varies from place to place. A few are directed at specific trouble spots due to drug peddling, prostitution, and other problems. Many others have a larger objective of preventing all crime. Several groups in the United States organize such patrols. The *Guardian Angels* are a prominent group that has a fair record of performance. They operate in more than sixty cities in the United States, Canada, and Mexico. A problem that is of some concern is the risk of vigilantism on the part of members of such organizations (Miller and Hess, 1994).

Citizen crime reporting encourages neighborhoods to organize themselves to fight and prevent crime. They gather at neighborhood meetings where they get to know one another and exchange information that has a bearing on crime. It is common for police officers to attend such meetings and provide information on the local crime situation and instruct citizens on how to request police assistance. Wide publicity is given for the existence of crime reporting programs so that they serve as a warning to possible offenders.

HISTORY OF CRIME PREVENTION IN INDIA

The report of the Indian Police Commission (1903) is the most prolific source on traditional crime prevention measures in the country. In the system inherited by the East India Company, which ruled parts of the Indian subcontinent from the Mughals in the seventeenth century, the primary responsibility for crime control in the rural areas was that of the Village Headman (VHM). The VHM was assisted by a few watchmen who were required to maintain a lookout for strangers, especially at nights, and to report on suspicious persons to the VHM. Interestingly, to supplement the efforts of the VHM, payments were frequently made to the leaders of plundering tribes so that they would prevent their followers from committing crime! One account (of Abul Fazul, Minister in the court of Mughal Emperor Akbar) that described the system under the Mughals, which was carried over by the East India Company said,

> A certain number of persons in each district shall be appointed to patrol by night the several streets and environs taking care that no strangers infest them, and especially exerting themselves to discover, pursue and apprehend robbers, thieves, cut-purses, etc. (India Police Commission, 1903:5).

Not satisfied with these arrangements, the company brought about a few reforms, including divesting landholders of their basic policing responsibility, thus cutting at the roots of the traditional village policing. The resultant situation was chaotic, and marked by increases in crime.

A committee, which was asked to look into the state of affairs, pleaded for the restoration of village policing because it was a system regarded as being in tune with rural culture, which evokes public cooperation for police tasks. This was readily agreed to by the court of the company, with some minor modification. Over the next few decades, several refinements were made, including the creation of a position of a Superintendent of Police, subsequently called Inspector-General. The Police Act of 1861 gave a formal shape to the system with suitable modifications. This Act is still in force, although the National Police Commission (1977) drafted a new Act. According to this draft, it shall be the duty of every police officer to "reduce the opportunities for the commission of crimes through preventive patrol and other prescribed police measures." It shall also be the officer's duty to aid and cooperate with other relevant agencies in implementing the prescribed measures for prevention of crimes. This explicit prescription was quite a contrast to the 1861 Act's laconic statement that one of the duties of every police officer was "to prevent the commission of offenses and public Nuisances."

The Indian Police Commission's (1903) report details the scheme now in vogue in the country for preventing crime. Two of its prominent features include surveillance of dangerous criminals and the dispatch of beat constables to villages. Although the former was introduced by the commission, the latter had already been in existence.

Even though habitual offenders (HOs) are supervised by the police, as prescribed by section 565 of the Criminal Procedure Code, other perpetrators are equally dangerous and should be under watch. As recommended by the Indian Police Commission, dossiers (or history sheets) are maintained for those individuals who seem inclined to lead a life of crime. Apart from HOs, usually two categories (known desperadoes and suspects) are known to the Indian police. These people are physically checked at periodic intervals by beat constables, and appropriate entries are made in their history sheets regarding their activities. This has proved to be a fairly effective crime control measure wherever dedicated personnel have performed the task of surveillance.

The beat system is another crime control device that historically has been used by grassroots police officers in different parts of India. This has been employed not only to keep an eye on offenders, but also to collect information on occurrences that have a bearing on crime. Instances are legion of beat constables visiting villages or mingling with the crowd in huge gatherings, identifying "bad elements" planning to commit crime and frustrating their plans. At the same time, it must be said that the beat system has not been exploited to its full potential, largely because of the failure of supervisory officers to prevent precious staff from being diverted to other routine work unrelated to crime control.

Certain states in the country have experimented with self-policing by communities. For instance, in the southern state of Karnataka, under the Village Defence Parties Act of 1964, the Superintendent of Police of every district establishes a party for each village (or group of villages) for guarding the village(s) and for preventive patrolling. A few modifications to the Act were carried out in 1975. The National Police Commission stated,

> In some districts where the local officers have shown interest and initiative, (they) have been quite active, but in several other places, the village public do not have a proper perception of the scheme and the scope for its utility to the community. The aspect of voluntary participation in village self-defense that is implied in this scheme has not been duly appreciated (1980:14).

Other known voluntary community groups organized with the objective of crime prevention were the Gram Rakshak Dals of Gujarat and Village Resistance groups of West Bengal.

The National Police Commission was positive in its 1980 report that crime prevention works, especially the operations against criminal gangs, which was best done by village defense parties rather than by individual village headmen. It added that the two together "should be viewed as a nucleus around which the concept of 'self-policing' by the community should be built and developed further." While projecting future needs, the commission, in its Final Report (1981), declared that one of the primary features of self-policing was "taking adequate preventive measures to protect life and property, and resisting an attempt on life and/or property should it take place in spite of preventive measures undertaken by exercising the right of private defence" (p. 23). In this context, the commission bemoaned the fact that most people did not take adequate preventive measures to safeguard their lives and properties:

> Proper locks are not used. A list of property in the house together with identification marks is seldom available. Often large amounts of jewelry and cash are kept at home instead of in lockers and banks (1981:23).

A disturbing feature in India is the fragile relationship between religious groups, especially between Hindus and Muslims. Indian history has been pockmarked by bloody riots. Police professional abilities to tackle riots objectively and fearlessly have been repeatedly questioned. But more than this, what has been assailed is the frequent police failure to enlist the cooperation of the warring groups to prevent such occurrences, or to bring the situation quickly under control once the conflagration starts. The civil administration normally resorts to the device of "peace committees" comprised of members of rival groups. The police play a significant role in the formation of such committees because of their intimate contact with the community. This point was succinctly brought out by the National Police Commission in its Sixth Report (1981:29):

> Prevention does not stop with increasing police presence or with immobilizing anti-social elements. A developing situation can be defused effectively by enlisting public cooperation. Peace Committees can play a very important role in removing fear, mitigating panic, reducing tension and restoring normalcy in the area.

EXAMPLES OF PAST PROGRAMS

The Delhi Police Experiments

Community policing programs in the past had focused attention on crime prevention. A few experiments conducted by the Delhi Police, an essentially urban police department serving a population of nearly 10 million, deserve a mention. In 1985, a scheme of Special Police Officers (SPOs) was introduced. These were able-bodied adults without a criminal record who were willing to undertake various tasks, such as patrolling, handling drug rehabilitation camps, training young girls in self-defense measures to ward off sexual harassment, and rendering help to victims of property crime. The success of this scheme could be gauged by the fact that more than 1,500 SPOs are now enrolled with the Delhi Police.

The Neighborhood Watch scheme launched in 1989 was aimed at reducing property crime and juvenile delinquency and bringing about better ties with the community. The police first identified a suitable community for motivating its residents to be alert in preventing crime. With the help of local SPOs and residents' associations, meetings were organized for the purpose of explaining details of the scheme. Thereafter, a Neighborhood Watch Committee was formed for each block of 500 residents. It undertook various anticrime tasks, including the strengthening of physical security arrangements in individual houses and registering the names of domestic servants and outsiders (such as vendors) who frequent the locality. Nearly 90 residential areas were said to have been brought under the scheme. The kind of motivation that had been aroused was evident from the fact that in one case of attempted burglary, after the occupant of the house sounded an alarm, nearly the entire neighborhood rushed to her rescue and successfully nabbed the culprit.

Another innovation was the "adoption" of a "high crime area." Apart from inducting additional police officers, the scheme involved special campaigns for the verification of the antecedents of domestic servants and for building public awareness on home security systems. This is known to have reduced crime appreciably, although exact figures are not available. Juvenile aid camps have also been set up by the Delhi Police to select willing street children, teaching them trades and later helping them find suitable jobs.

The Mohalla Committees of Maharashtra

A scheme to promote religious harmony, floated in 1990 by the police in Bhiwandi (Maharashtra State), a town notorious for Hindu-Muslim strife, can rightly be regarded as an attempt at community policing. Local bodies called Mohalla Committees were formed to sort out problems in the community. About thirty members in each committee used to meet once a week and discuss various issues with the local Sub-Inspector of Police. It was agreed that the police would help the community even in nonpolice matters, such as the supply of electricity and availability of rationed food commodities. The close relationship the police forged with the population ensured that the ground level intelligence required for monitoring the state of Hindu-Muslim relations was available without any extra effort. Further, thanks to the increased opportunity for Hindus and Muslims to interact among themselves through the Mohalla Committees, violent exchanges had become few and far between.

The Karnataka DGP's Directive

The Karnataka State Police, in the south of the country, also has been making efforts to take the message of community policing down to the police station level. A Standing Order, issued by the DGP in February 1994, directed the formation of a citizens committee to meet at least once a month to discuss matters such as the deviant behavior of individuals, patrolling, and distribution of protective duties during outbursts of crime. Further, police officers visiting villages on their beat were directed to ascertain citizen grievances, not only with regard to the police, but also in matters such as street lighting, water and power supply, conditions of roads, and sanctions of old-age pension. The DGP's order contemplated the grant of incentives to police officers who achieved significant results in the area of police-public relations. No material, however, is available as yet to measure the extent of implementation of this order and its impact on the average citizen.

The Tamil Nadu Program

Policing in the southern state of Tamil Nadu has been traditionally known for its emphasis on good relations with the public. A forum in the form of a Village Vigilance Committee (VVC) at the police station level is available for promoting this situation. Viewed mainly as a crime prevention measure, the VVC's effectiveness has depended almost wholly on the enthusiasm displayed by the individual station house officer. It is known that, over a course of time, VVC meetings have become a mere ritual intended for the consumption of supervisory officers during their inspection of police stations.

SOME RECENT PROGRAMS

Alongside the distressing apathy of field officers in essentially rural settings, one sees a new awareness among higher police officers in the larger cities of the need to involve citizens in police chores. This trend is the direct result of the growing complexities of urban policing, especially the increase in violent property crime and its extensive reporting by the print media. There is some recognition by the top brass that unless better communication with the public is established and the latter are directly involved in policing tasks, the police force is likely to become overwhelmed and even paralyzed by the staggering demands of order maintenance and crime prevention.

A very recent innovation in the area of community policing was the establishment of the Friends of Police (FOP) in the southern state of Tamil Nadu. Appalled by the widespread negative image of the police in the eyes of the public, Prateep Philip (1998) launched the FOP movement to "foster the hitherto untapped wellspring of public goodwill for the police." In the

Ramanathapuram District, where it was introduced for the first time a few years ago, 100 volunteers from different walks of life responded immediately. Philip (1998) claims that membership has swelled to nearly 100,000 and that the movement has extended itself to many other parts of the state.

One major FOP objective is "to promote crime awareness within the community and foster crime prevention [and] to act as a force multiplier by involving FOP members in community policing, and thus increase police potential to deter and prevent crimes" (Philip, 1998). Success has come in several forms, including information that led to the identifying of an offender involved in a case of homicide. In another case, FOP members who were fishermen supplied information that resulted in the seizure of explosives. A recent survey revealed that the FOP concept was considered useful both by the police and the FOP members. There is, however, the feeling that many more Tamil Nadu supervisory officers need to be convinced of the feasibility of the FOP.

The Chengalpattu East Police District, on the outskirts of Madras City, is another area where anticrime experiments have been tried out with the cooperation of the police. For instance, a scheme has been devised to combat late-night raids on petrol stations by criminal gangs. There are 83 such outlets in the district which have become quite vulnerable. At a recent meeting convened by the Deputy Inspector General of the area, owners of the petrol stations agreed to erect collapsible iron gates to strengthen the cashier's room and to install alarms. Additionally, they will use security guards and watch dogs to ward off possible offenders. It was further agreed that no customer would be allowed to get into the cashier's cabin after 6 p.m. All transactions, thereafter, will be conducted through a protected counter. This is one instance of the police taking the initiative in crime prevention activity.

Similar programs have been drawn up and executed in the Chengalpattu East District to counter gang robberies. The emphasis is on encouraging the public to collect and pass on information regarding movements of strangers and to arm themselves with long sticks (*lathis*) as a measure of self-defense. Motorcycle patrolling has been organized to check gangs that move about on motorcycles, snatching jewelry from unwary women.

There is undoubtedly an awareness that there is no alternative to going into the community to seek greater involvement of the public in the daily chores of the police. There are two factors, however, that have operated against success in the endeavor to draw in the community. The first is a declining focus on local community problems. This decline is explained by the increased attention paid to escalating group violence which transcends the local community. Perhaps a more important limitation is the fact that the average constable is ill-equipped by training, law, and police tradition to interact with the citizen at a level of mutual confidence. These constraints are not likely to disappear for a long time.[1]

CHALLENGES AND PROBLEMS

Although crime prevention as a strategy is of great appeal to the police and the community at large, the efforts in India to bolster such a strategy have been only a little more than modest. A variety of reasons may be cited to explain this apparent apathy. First, it is a fact that no state police force in the country would like to admit that crime is on the increase. Even when statistics point to a rising trend, the tendency is to wish it away as a temporary phenomenon that deserves to be ignored. It is not surprising that, in such an atmosphere, there is no great urge to talk about crime prevention.

Second, there have been very few field studies in India, either by scholars or police officers, to investigate the impact of the small number of crime prevention projects. In the absence of such studies, the message of crime prevention is not disseminated to social groups who could be expected to take up its cause. Third, in order to succeed, a crime prevention strategy needs a solid database, especially in terms of victims' experiences. In spite of the fact that the Indian police have made impressive strides in computerization, not enough attention has been given to the need to collate information on a scientific basis so as to support experiments.

A final reason why crime prevention has not taken off in a big way is that policing remains insulated from the operations of local and civic administration. The mayor's administration in big cities, and its counterpart in smaller towns and villages, has no say in policing matters. As a result, the two authorities operate in isolation. They do not realize that, acting in tandem, they could tackle each other's problems in a much more effective manner than at present. Street lighting, removal of garbage, provision of uninterrupted water supply, eviction of encroachments of sidewalks, and similar problems have a direct and indirect impact on crime. Coordination between the police and city authorities in these spheres is invariably left to the whim and caprice of individual officials, and it is not mandatory for them to work together to resolve issues.

The Legal Difficulties

Section 149 of India's Criminal Procedure Code (Cr.P.C.) states that

> Every police officer may interpose for the purpose of preventing, and shall, to the best of his ability, prevent the commission of any cognizable offence.

Section 151 of the same code empowers the police to make arrests for the prevention of a cognizable offense. This is invariably used to frustrate the designs of groups or individuals to disturb public peace.[2]

Two other provisions are also of great help. Under Section 107, persons likely to cause a breach of the peace can be bound over by a magistrate to keep the peace for a year. Any violation of the conditions imposed invites a term of imprisonment. This provision is invoked frequently by the police to defuse situations in which individuals or groups are locked in a local dispute (usually over ownership of land or right of worship in a religious place) that threatens to disturb peace.

Section 109 helps to bind over to good behavior (for a maximum period of 1 year) any person concealing his presence with a view to committing a cognizable offense. This provision is very often employed against property criminals who display signs of recidivism. Perhaps more useful is Section 110 that is directed against habitual offenders. A wide range of offenses, including the receiving of stolen property and harboring of criminals, and hoarding, profiteering, or adulterating food or drugs, come under the purview of this section. Also, unlike Sections 107 and 109, the bond executed by an accused under Section 110 is valid for 3 years.[3]

The dileneations above are no doubt impressive on paper, but they brim with practical difficulties. For instance, a person arrested under Section 151 cannot be detained in custody for longer than 24 hours unless his further detention is required or authorized under some other provisions of the Cr.P.C. or any other law. This is a handicap that considerably dilutes police capacity to act against miscreants who are ingenious enough to indulge in activities that are difficult to be brought under other law.

Also, action under Sections 107 and 109 is prolonged, and several hearings in court precede any final orders binding a person to maintaining peace and good behavior. Further, magistrates are extremely circumspect in passing orders against individuals arraigned by the police. A Supreme Court of India pronouncement on the scope of Section 107 Cr.P.C. is illustrative of the situation:

> [The] object is prevention and not penal, as the matter affects the liberty of a person who has not been found guilty of an offence. Power should be exercised by Magistrates in accordance with law. If the court finds that no untoward incident happened from the date of incident complained of and a long period has elapsed, the court may draw the inference that danger of peace has vanished (*Ram Narain Singh v. State of Bihar*, AIR 1972 SC 2225).

THE FUTURE

There is evidence of a growing interest in the subject of crime prevention on the part of the community at large. Instances of groups, both organized and unorganized, wanting to take the initiative are being reported from different parts of the country. Interestingly, many of them are independent of police efforts. This is explainable by an unconcealed cynicism over police unwillingness to assist in crime that may not be grave, but which, nevertheless, is a source of great annoyance to a neighborhood. This phenomenon is frequently seen in big cities where independent houses are giving way to blocks of multistoried apartments. Residents here are acutely conscious of security needs and hire private guards to keep an eye on what goes on in the neighborhood and to strictly regulate vehicles in the vicinity and visitors to such apartments. The incredible expansion of shopping malls and hotels is another development that has prompted an increased demand for nonpolice private security arrangements. Uniformed security personnel and closed-circuit television cameras are now an ubiquitous presence.

These trends clearly point to an increasing public awareness of the need to tighten security in the working and living environment. What is striking is the subdued role played by the police and other governmental agencies. This is no doubt in tune with the growing emphasis on privatization in all spheres of life. Nevertheless, a dominant role will have to be played by the police in educating the community on the basics of crime prevention. Without such a role, one cannot see crime prevention strategies breaking new ground in terms of their sweep and reach. As it is, there is no evidence that the police leadership is aware of its responsibility in this important sphere.

ENDNOTES

1. We would like to acknowledge that this perceptive analysis of the scene is that of Mr. N. Krishnaswami, retired IGP, Tamil Nadu, who recorded this while reviewing the draft chapter on community policing in Raghavan's forthcoming book *Policing a Democracy: The Case of India and the U.S.* (New Delhi: Manohar).
2. The National Police Commission (1977) suggested amendment of Section 151 Cr.P.C. with a view to providing for the preventive detention of potential criminals, for a short period, say, of one or two weeks. At present, there is no law that permits such short-term detention. The National Security Act and the Goondas Preventive Detention Act contemplate only longer detention.
3. Police officers employ all three sections in their routine with a view to maintaining peace and containing crime, especially property offenses. It is the experience that some officers at the police station level resort to such action mechanically and too frequently to boost statistics and impress supervisors.

CHAPTER 9

THE KENYAN PERSPECTIVE ON COMMUNITY POLICING AND CRIME PREVENTION

Mary M. Mwangangi[1]

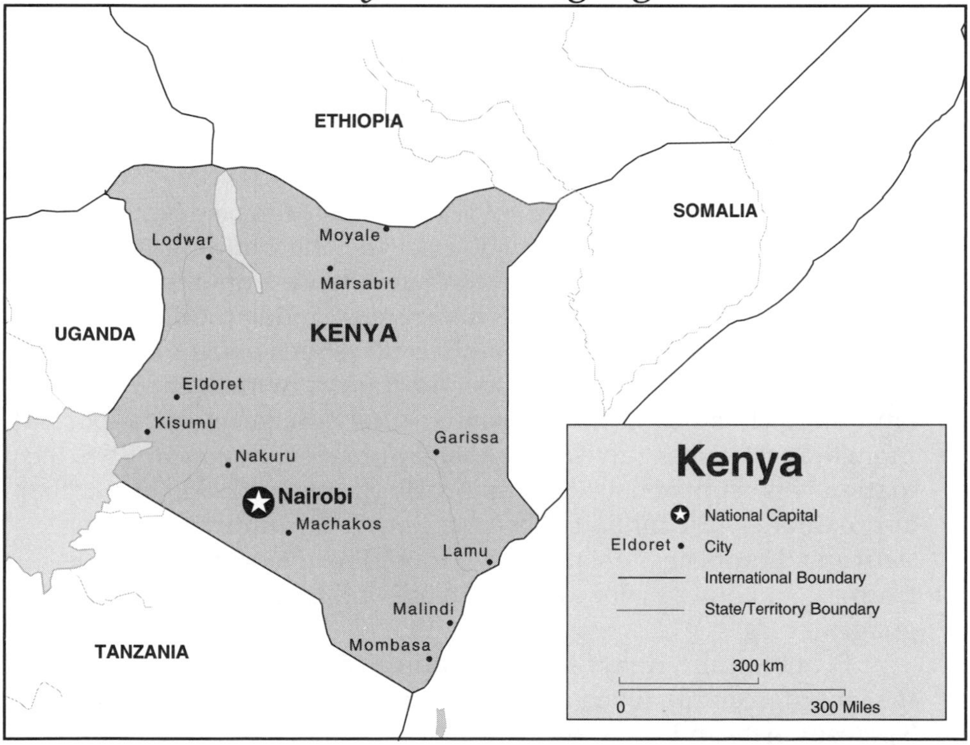

The universal thrust of crime prevention revolves around the control of crime. In precolonial times, indigenous Kenyan communities did not have a police force. Individual communities regulated the behavior of their members by means of age-old traditions. The central authority exercised by a council of elders had the power to arbitrate cases and impose punishments.

The approach to crime control in Kenya, especially in the first half of this century, was largely reactive. Policing involved the committing of a crime and the subsequent investigation into its solution. This system of policing, also known as the "fire brigade" model, has undergone tremendous change since 1963. Significant innovations seek to include members of the public as equal partners in crime prevention. This chapter discusses these innovations in areas of crime prevention and community policing to control drug trafficking, possession of illegal firearms, motor vehicle theft, and fraud.

COUNTRY PROFILE

Kenya lies across the equator on the eastern seaboard of Africa. Its land mass is approximately 583,000 square kilometers (224,961 square miles), slightly larger than France. Kenya gained independence in 1963, after having been a British colony since 1895. The British annexed Kenya because of her strategic position in the region during the building of the Uganda railway (1895–1901). On attaining independence from the British in 1963, Kenya adopted a western-style democracy under the Lancaster House Constitution. This provided for an Executive President, a Parliament, and the Judiciary. Kenya is basically an arid country, with over 80 percent of its land mass classified as arid or semiarid. The majority of the population, more than 80 percent, lives on the remainder of the land. In 1996, the population was estimated at 28.3 million. By the end of 1998, the number was approximately 30 million. The population is multiracial, composed of Africans, Europeans, Asians, and Arabs. The majority live in the capital, Nairobi, and other major towns. Swahili and English are the primary languages.

Economically, Kenya is an agricultural country. Sixty-seven percent of the labor force is involved in agriculture, 21 percent in the informal sector, 3 percent in the industrial sector, and approximately 9 percent in the formal (wage-employed) services sector. The national gross domestic product in 1997 was estimated at US$10 billion. Over the period 1990–1997, the Kenyan economy witnessed an average growth of barely 2.8 percent per year, not enough to increase the standard of living. The slow growth has been due to factors such as drought, the El-Niño phenomenon, reduced foreign assistance, and high domestic interest rates.

THE JUSTICE AND POLICE SYSTEMS

Police work in Kenya can be traced to the country's colonial past. Despite this relatively long history, very little has been written about the Kenya Police Force or about matters relating to crime prevention. Confidentiality and poor police-public relations have discouraged researchers from investigating issues concerning the police and their role in crime prevention. Foran (1962) focuses on the activities of the Kenya police before independence. Between 1887 and 1918, under the British, policing and crime prevention was something of a paramilitary effort. When the country evolved from a protectorate into a colony in 1920, the police force became known as the Kenya Police Force. The force reorganized, changed responsibilities, and initiated the Kenya Police College. The establishment of the Police College marked the beginning of a conscious and deliberate effort to train officers in crime prevention skills.

With the advent of independence, crime prevention in Kenya captured the imagination of scholars and researchers. Mushanga (1976) states that the best way to deal with crime is to prevent it from happening. Mushanga further argues that it becomes very expensive to deal with the consequences of crime. Costs include the resources used in investigation, apprehension, prosecution, and imprisonment of offenders. Mushanga (1976) also argues that in order to understand crime, we must study crime and its causes. He insists that law enforcement must be matched by trained personnel in psychology, sociology, law, and politics. The police must, of necessity, be trained in these disciplines.

Awuondo (1996) believes that the way forward is to improve prosecution through careful investigation. Positive results will motivate the public to volunteer information and help rid society of dangerous criminals. Law enforcement agencies should function as part of the wider society and not as isolated self-sufficient entities suspended above society. He further states that the public is highly segmented. He calls for a proactive, people-oriented policing approach where law enforcement regards everyone equally as the public, protects the public, and works for and with the public.

Muga (1972) claims that crime is a result of poverty and urbanization. Rural to urban migration in search of employment is problematic. Society should complement the work of the police and the prison department with crime prevention techniques and rehabilitation of criminals. The education system can play an important role by teaching good citizenship. At the same time, churches and other social welfare organizations should be involved in community-based counseling. In rural areas, communities can tap the accumulated wisdom of their elders to help guide deviants. Finally, Okola (1996) contends that victims of crime contribute to its commission by exposing themselves and creating conditions that make crime possible (for example, leaving precious goods in a car).

The justice and police systems of Kenya are based on the assumption that the majority (about 80 percent) of the country's population and visiting residents are law-abiding and peaceful. The minority (20 percent) are either hard-core offenders or persons who engage in criminal activities of a petty nature. In recognition of this, the justice and police systems are designed to check the activities of the criminal minority and to protect the rights of the law-abiding majority.

The Kenyan justice and police systems have put in place an elaborate arrangement of checks and balances to ensure that the rights of suspects are upheld and that they are treated fairly. This has been realized through a separation of roles. The investigation and arrest phases have been left to the Kenya Police Force. Prosecution for complex cases (for example, murder and treason) has been assigned to the Attorney General and the Director of Public Prosecutions. For ordinary cases, the Police Force has specially trained personnel who prosecute cases in magistrate's courts. All cases, unless under very special circumstances, are held in open court.

As in many other countries, the prosecution is required to prove the guilt of an accused person beyond reasonable doubt. In the Kenya justice system, an accused person is presumed innocent until proved guilty in a court of law. Once a person is found guilty and is given a custodial sentence, the system has assigned the Kenya Prison Service the task of carrying out the sentence. For nonfine, noncustodial sentences, a National Probation Service handles the role of rehabilitating the convict. Although it is generally expected that the prison or probation service will rehabilitate convicts, overcrowding in prisons hampers this objective. For those under age 18, Borstal (reformatory) institutions and approved schools substitute for prison service. In addition to rehabilitating the youths, they provide formal and vocational education. A guiding principle behind punishing crime is to deter other people from committing crime.

The Goal, Philosophy, and History of Crime Prevention

Historically, the universal goal of crime prevention revolved around the control of crime. In the precolonial era, indigenous Kenyan communities did not have a police force. Individual communities observed age-old mores and traditions of respect for the sanctity of human life and communal property to regulate the behavior of their members. Customs and taboos were employed to guide community members in their interpersonal relationships. These traditions, coupled with the central authority provided by a council of specially selected elders, served as a guide in arbitrating cases and meting out punishment to the offenders.

The novel idea of a standing police force was introduced with the advent of colonialism. The Imperial British East Africa Company, to whom the idea of a standing force in Kenya can be traced, formed a band of ex-military men to guard the company's property in Mombasa. In 1920, Kenya changed its status from a protectorate to a colony. The police force also adopted its present title (the Kenya Police Force) and moved its headquarters from Mombasa to Nairobi. The force expanded rather quickly, partly influenced by the influx of settlers who were either discharged or retired from the British Army. The force was run along the lines of the British Metropolitan Police. When Kenya attained its independence in 1963, the force appointed indigenous Kenyans to most of the senior positions, including that of the Commissioner of Police. Since then, the force has grown tremendously in terms of size and operation.

Crime control in Kenya, especially in the first half of this century, was largely reactive. This model depended on the commission of a crime and its subsequent investigation. As cited earlier, this system of policing, known as the "fire brigade" model, has undergone tremendous change since independence. Significant innovations have sought to include members of the public as equal partners in crime prevention. Over the last four decades, the Kenya Police have evolved a proactive policing model that does not wait for a crime to be reported before action is taken.

This community-oriented policing involves working with citizens to find solutions to community problems. It involves leaving patrol cars behind, reaching out to the people, and identifying problems. It establishes meaningful rapport with the public and then endeavors to educate citizens about their important role in law enforcement. The community policing concept empowers citizens by inviting them to form security committees and citizens' patrols. Citizen input in policing, as well as contributions of money, materials, and time, eases the burden on the police force. This model, hopefully, will carry Kenya successfully through this new millennium.

Existing Police Crime Prevention Projects

Crime prevention is paramount to successful policing. Prevention of crime implies instituting mechanisms to ward off potential law breakers, either in the planning or execution stages. In Kenya, the most troublesome crimes are those related to drug trafficking, car jacking, and illegal possession of firearms. Each of these is discussed below.

One new program that can be of significant benefit in solving crime is the police hotline. The hotline system was introduced in Nairobi in 1995 and has been extended countrywide. The hotline is a direct telephone line to a police unit, which is on duty 24 hours a day. The aim is to ensure that serious crime reports receive immediate attention. Another advantage of the hotline is that the person making the report need not reveal any personal identity.

The hotline was introduced when it was realized that members of the public did not wish to be directly involved with the police. Consequently, citizens would watch a crime take place and fail to report it. The hotline has been very effective. Recently, for example, it was instrumental in the arrest of a suspect who had stolen US$1 million from the Jomo Kenyatta International Airport.

The hotline, which is still in the experimental stages, is limited to the major towns and to people who have access to a telephone. Another challenge is that some callers take advantage of the anonymity and make false reports, often resulting in futile investigations. Information received on the hotline has also been challenged in court when used as evidence, since the source of the information may not be traced. Similarly, those who provide false or alarming information (such as a bomb hoax) cannot be traced for prosecution.

Drug Control

Increasingly, Kenya is becoming a transit point for drug traffickers. For example, mandrax and cocaine find their way into the country from India and South America. In addition, Kenya produces *Cannabis sativa,* a plant grown in virtually all corners of the republic. The Kenyan government is cooperating with other governments in training law enforcement personnel to curb the drug menace. In the meantime, the Kenya Police Force has taken measures to check drug trafficking and use.

The Anti-Narcotics Unit has set up an intelligence unit whose function is to collect tactical, strategic, and operational intelligence on drug crimes. Raids are a key operational technique, whereby sporadic visits are made by police to disrupt criminal activity. Apprehension rates are low, but this method causes uncertainty for criminal gangs.

Unfortunately, the drug problem cannot be solved by the Anti-Narcotic Unit or the police force alone. The incorporation of other enforcement agencies, such as the Customs and Immigration Department, has been encouraged. In the planning stages is the signing of memoranda of understanding with private companies—such as shipping agents, clearing and forwarding companies, airlines, travel agents, and courier companies—to assist in obtaining information on clients. The Kenya Anti-Narcotic Unit also cooperates with international drug control organizations such as the United Nations Drug Control Program and the British Drug Liaison Services.

The media have also been important players in creating drug awareness among Kenya citizens. Radio coverage has offered programs sponsored by both government and nongovernmental organizations to warn people about the dangers of drug use, reaching out especially to those in rural communities. Television has been instrumental in featuring educational programs that focus on the evils of drug abuse. These programs highlight various aspects of addiction to help produce a drug-literate society that can say no to drugs.

The use of "alternative media" is peculiar to the third world countries where the broadcast and print media networks may not be as developed as they are in western countries. Alternative media educate illiterate or semiliterate masses in rural communities through activities such as traditional dances, drama, and storytelling (lectures) to convey important messages on problems affecting the society. The Provincial Administration organizes public meetings, called *barazas,* with the help of professionals to convey messages on drug control. This is done both in schools and villages.

Though there is provision for establishing rehabilitation centers under the Narcotics Act, such centers have not yet been set up. Nongovernmental and religious organizations involved with rehabilitation of street children play a major role in reducing potential drug abuse. Street children often abuse drugs in order to commit crimes. Nongovernmental organizations are encouraged to support rehabilitation programs to reduce drug demand by young people.

Control of Illegal Firearms

The geopolitics of the East African region has a direct bearing on the influx of illegal firearms into the Kenyan territory. With the change of government in Addis Ababa, the collapse of the Somalia government, and the ongoing conflict in southern Sudan, the two major borders (northern and eastern) have become unstable. The number of small arms originally in the possession of the military, police, and the paramilitary in these countries fell into the hands of factions or individuals with little or no interest in group security, other than individual self-protection. The influx of these firearms through Kenya's borders can be attributed to the economic value of the weapons. The habitation of some tribes across these common borders presents a challenge for police, while it makes it easier for arms traffickers.

The Kenya Police Force has a special unit charged with the control of civilian and police firearms. The Central Firearms Bureau employs various methods to control the flow of arms across common borders. These include disarming, amnesty, and control of entry/exit (border) points. To disarm those who possess illegal firearms, sporadic raids are carried out in targeted areas, and people found in possession of such weapons are arrested and prosecuted. Amnesty does not require enormous financial resources, and it allows those who are holding illegal firearms to surrender them to government officers. Finally, police and customs officials are vigilant at all official entry points and airports.

Among the crime prevention methods, not sponsored by the police, other government agencies, like the Provincial Administration, have been instrumental. The Provincial Administration conducts raids to confiscate illegal firearms. The advantage of using chiefs and district officers in the process is that they are in close touch with the communities they administer. Hence, they have developed an extensive intelligence network. All arms recovered in this manner are surrendered to the police force. The Kenya Wildlife Society, a quasi-governmental agency charged with combating poaching activities in national parks, has also played a vital role in the recovery of illegal firearms. It patrols the national parks and arrests poachers who later face criminal charges.

Community input comes in the form of home guards—groups of citizens who volunteer their services in the protection of their neighborhoods. They are specially trained in the use of firearms and deployed mostly in areas where banditry and cattle rustling is rampant. Firearms recovered as a result of their efforts are duly handed over to the Kenya Police Force.

Motor Vehicle Theft

Traffic in stolen motor vehicles continues to be a serious security issue within the country and the entire East African region. Theft of motor vehicles is influenced by several factors. First, the high rate of unemployment means that young men and women are often unable to earn an honest living. Motor vehicles become particularly attractive targets for those who wish to make quick money. Second, the increased unit cost of motor vehicles has forced people with inadequate finances to commit theft. Third, the increased cost of spare parts also contributes to the rate of motor vehicle theft. The vehicles are cannibalized, and the spare parts are offered for sale. The net effect of this particular crime is great economic loss to the public, insurance companies, and the three countries in the region.

The police attack this crime using varying techniques. The flying squad is a patrol unit within the Kenya Police Force charged with the responsibility of collecting intelligence on car theft and robberies. Rural surveillance is coordinated by foot and car patrol. These patrols have arrested suspects who, on interrogation, turn out to be wanted car thieves or robbers. Design surveillance is adopted on the basis of crime data analysis. When a particular area is identified as prone to robberies and carjackings, officers are deployed to those places. Officers are also sent to areas outside Nairobi for surveillance and collection of criminal intelligence. This approach is aimed at stopping stolen vehicles from being driven across the borders.

Whenever a vehicle is reported stolen, the police circulate the information to all stations in the country. Using a description of the stolen vehicle, patrol teams track the vehicle. This may include setting up roadblocks on routes where the vehicle is likely to be taken. Police also visit suspected hideouts. The unit also has established a network of informers who provide information on carjackers and motor vehicle thieves. Such information assists in the recovery of the stolen vehicles.

Several handicaps are faced in the course of attacking motor vehicle theft. One is inadequate communication equipment, which reduces the extent to which information is circulated to police officers. Second, the standard patrol vehicles in use are not built to carry out extended surveillance along the borders. Such surveillance requires four-wheel-drive vehicles. Third, members of the public are not well informed on what precautions to take to protect their vehicles. There also is strong suspicion that some of the theft reports are false. False reports are given for the purpose of defrauding insurance companies. Fourth, reports of theft often do not reach the police in good time. This makes it difficult for the police to track the vehicles because reports are received after the vehicles have already reached the intended hideouts or have already been driven across the border.

POLICE-COMMUNITY PARTNERSHIP INITIATIVES

Beyond activities aimed at specific crimes, the police are involved in more general prevention efforts. The police-community partnership concept, adopted by the Kenya Police under the broad umbrella of "community policing," has been borrowed from the developed world. This concept establishes bonds between law enforcement agencies and the communities they serve. This partnership is essential for achieving significant reductions in crime.

However, the concept has had to be modified because Kenya is a developing nation with a low per-capita income and a fragile economy. As a result, various communities have started making monetary and material contributions to law enforcement. They do this by assisting in the construction of police stations, providing patrol vehicles, and patrolling their own areas. This allows the Kenya Police Force to focus its meager resources on needy communities that cannot emulate their richer counterparts. Thus, Kenya's concept of police-community partnership differs from that of the developed countries. The following are examples of ways in which Kenya has modified the concept to suit local situations.

Parklands Neighborhood Watch

The Parklands Neighborhood Watch Project (NWP) was started in 1992. The largely affluent Asian community in the area felt threatened by the rising levels of crime. Police administrators sought to engender innovative crime prevention methods with the assistance of the residents. The residents provided personal motor vehicles, and the Kenya Police Force provided armed personnel. This marked the birth of Parklands Neighborhood Watch Project.

The NWP provides at least four cars with drivers every day. The police provide two armed officers for each vehicle. The vehicles are assigned separate patrol areas. Whenever there is need, the neighborhood watch increases the number of patrol vehicles. The NWP has established an operations room in a local school where all messages are received and transmitted. The NWP is divided into six departments. These are the trustees and board of directors, the executives running the organization, the public relations officers, the transport officials, an emergency response unit, and controllers who supervise the control room.

Occasionally, members fail to provide vehicles. Similarly, due to exigencies of duty, police officers may be unavailable. Such problems, however, are temporary in nature and are addressed during regular monthly meetings. Due to racial tension and socioeconomic disparities between the affluent organizers of the NWP and other residents in the area, there is also a feeling that the NWP is elitist and ignores the crime problems of the less affluent, leaving the policing of the less fortunate people to the regular officers only.

The NWP has achieved significant successes in the reduction of crime and increased public confidence in the local police. It is a project that should be supported and maintained. Indeed, it has eased the burden of policing for the local stations by providing transportation and communication equipment.

Similar projects are in operation in other districts with large rural populations. The business community in the Machakos and Kitui Police Station areas have similar, but smaller, organizations. These organizations provide a vehicle and driver, and the police provide armed officers.

Hardy Community Police Project

The Hardy Police Station is situated 8 kilometers from the Nairobi city center. The post was established in 1953 to serve a small community of colonial settlers. Due to the growing population in the area and increases in crime, the community felt that there was need for a bigger, properly manned police station. In 1990, the residents of the area formed the Karen and Langata District Association, whose purpose was to improve the security needs of the area, among other things. In this regard they resolved to build a police station. They approached the commissioner of police to allow them to contribute toward improving the police post to full police station status with the assistance of the government. The commissioner agreed to the proposal.

The residents organized themselves and contributed money and materials for building the physical structures, and the police force provided vehicles and professional advice for the construction of the building. Phase one of the project is complete, with modern cells, a kitchen, and dog kennels. The building was officially handed over to the commissioner of police on May 26, 1998. The police post has since been upgraded to a fully fledged police station. Phases two and three of construction work are still in progress with significant citizen input. The success of the Hardy Police Station project is a milestone in citizen-police cooperation in crime reduction.

Viwanda Location Community-Based Crime Prevention Project

This project in the eastern part of Nairobi city was not initiated by the police. The project operates in a large slum area, divided into seven administrative units, each with a chairperson chosen by the people. The security system in these slums is organized by the area administrator (chief) together with the residents. Each of the seven chairpersons helps the chief deal with security matters in their respective areas.

The chairpeople supervise between twenty and forty people (called vigilantes) who patrol their areas during the night, with the assistance of two armed police officers. The vigilantes are residents of the villages who are either self-employed or work in nearby factories. They offer their services on a voluntary basis. Since they reside within the slums, they are familiar with the environment and with the people who live there. The vigilante groups also collect intelligence on wanted or would-be criminals, which they pass on to the local police station.

The vigilante team also assists in other nonsecurity matters, such as natural disasters. During such times the vigilante groups come to the rescue of those affected. The volunteers are screened by the chief with the help of the area police boss. This helps eliminate those who might have been involved in past criminal activities. Only volunteers with clean records are enlisted.

This vigilante security system has several merits. First, police efforts in curbing crime in the slums have been boosted, with more crime being detected. As a result, crime rates in the slums have gone down by about 60 percent. Second, the slum dwellers are eager to participate in security matters and are directly involved in managing the security of their locality. Third, in the past, the residents used to take the law into their own hands and dispense "mob justice" on suspected criminals. A number of these suspects were stoned to death or burnt using a lynching method. With the introduction of the vigilante system, suspects are now promptly handed over to the police and mob justice has been reduced.

At the same time, this system has its drawbacks. Volunteers who stay long in the vigilante security system tend to get involved with criminals and at times are compromised. Others tend to lose interest due to the fact that there is no monetary gain for their effort. Also, well-to-do criminals have been known to use money to bribe volunteers in withholding important information that can lead to arrests.

The Kenya Banking Fraternity–Police Partnership

A final program in which police are heavily involved deals with bank fraud. As a result of mounting white-collar crime, the Kenya Banking Fraternity, under the umbrella of the Kenya Bankers Association, resolved to cooperate with the police force to find common solutions to threats on the economy from fraud offenders. The bankers offer financial support for specialized training of fraud investigators at the Criminal Investigation Department (CID). The bulk of the trained detectives are then deployed in the CID Fraud Squad and the Central Bank of Kenya. As the program is still in its formative stages, its success has yet to be measured. However, it has been generally accepted that the quality of investigation has improved.

CRIME PREVENTION INITIATIVES NOT SPONSORED BY THE POLICE

There are numerous initiatives to fight crime that are not sponsored by the police. Two such programs are discussed here. First, security firms have introduced a new service whereby they install radio alarms in commercial or industrial and residential premises. The firms also have placed squads of guards in motor vehicles at strategic points close to the premises. When the alarms are activated, the squads swing into action while the police are alerted for backup.

A second approach is the control of private taxis (*matatu*). These are licensed public service vehicles on Kenyan roads which supplement bus services in the country. This is a result of a rapidly growing population that has outstretched the regular bus services. Inaugurated two decades ago in a haphazard manner, this mode of transport has evolved into a vibrant industry compelling the special attention of the Traffic Police Department. The *matatu* industry has been plagued by petty crime committed against commuters, often by their employees. This behavior has also led to serious infringements of traffic regulations, resulting in the loss of lives through road accidents. This problem is particularly acute in urban areas.

The Nairobi Bus Terminus *matatu* organization is an example of a police-community partnership in crime prevention and traffic control. This organization has formed a security network whose duty is to oversee the smooth running and security of the commuters at the terminus. The purpose is to facilitate the smooth flow of traffic in and out of the terminus and to check on and control petty crimes like pickpocketing and mugging. The citizen volunteer security details are screened by the owners of the motor vehicles, with the assistance of the police, before they are assigned to their duties. They are also issued identification cards or badges so that commuters can identify them easily. This kind of privately initiated project has helped minimize petty crime at bus stops and terminals within Nairobi city and other towns across the country.

FUTURE OF CRIME PREVENTION

The challenges to the success of crime prevention in Kenya are multiple. As stated earlier, the country has been experiencing low economic growth. This means that more people will be unemployed or will have no means of livelihood. As a result, crimes such as theft, robbery, and fraud are expected to increase. The unemployed also may be tempted to use drugs. Kenya however, is preparing to meet these challenges through innovative preventive measures and police-community partnerships.

ENDNOTE

1. The writer acknowledges the support she received from the Anti-Car Theft Unit, the Firearms Bureau, the Anti-Narcotics Unit, and the Banking Fraud Unit, and from Chief Inspector Kibunja, Inspector Wambilyanga, and Inspector Rukungu of Police Headquarters—Nairobi. The efforts of Ms. Rachel Ratemo and Ms. Benedictar Kariuki who typed this chapter are also greatly appreciated.

CHAPTER 10

CRIME PREVENTION IN NIGERIA: PRECOLONIAL AND POSTCOLONIAL DIMENSIONS

Obi N.I. Ebbe

Crime is committed in every healthy society. As Emile Durkheim (1933) put it, even in a society of angels, a common sin would become a crime. And Nigeria is no exception. Crime has always been present in Nigeria, and it has provoked moral outrage in both precolonial and postcolonial times. In precolonial Nigeria, however, crime was so abhorred that it created two classes of people: the Nobles (*Nwaafor*) and the Outcastes (*Osu*) among the Igbos (*Ibos*), one of the peoples of Nigeria.

People who committed an abomination, or serious public offense, (*aru*), such as the murder of one's parents, brother, sister, or a relative; incest; or any offense against the gods, were assumed to be dedicated to the god of a strange land, and they became outcastes, or untouchables (*Osu*) (Ebbe, 1996). Private offenses such as burglary, robbery, and stealing were disposed, not by imprisonment, but by restitution, compensation, communion (or reconciliation) feast, or fine.

Throughout precolonial times, crime prevention and control were the exclusive responsibility of the head of a household. The male head of a household had the prerogative to grant mercy for any offense. Another major responsibility was the preservation of life and the protection of the property of each of his dependents. This latter responsibility was always kept in mind when the residential home was designed.

The central focus of this chapter is to investigate the patterns of social organization for crime prevention and the strategies designed by the national government, local governments, and the village councils for crime prevention and control in precolonial and postcolonial Nigeria. It is beneficial to know what these old strategies were and why they were abandoned, the nature of the new strategies and how they differ from those in the past, how they can be distinguished from crime prevention efforts in other countries, and what other countries could learn from Nigeria's crime prevention attempts. A great deal can be learned when patterns of crime prevention are studied historically and cross-culturally. In doing so, one can see how wars and interethnic conflicts influence strategies in crime prevention.

EARLIER STUDIES ON CRIME IN NIGERIA

There is no published text on crime prevention in Nigeria. Studies of Nigerian criminology have concentrated on the etiologies of crime and delinquency, police administration and operations, and correction methods. Okonkwo (1966) and Tamuno (1970) pay little attention to crime prevention in their books. Both authors, writing in different decades about the Nigerian police, have reported only the urban-based police watch and patrol systems of crime prevention. Neither presented nor discussed the traditional methods of crime prevention and the private citizens' physical and environmental security design systems for crime prevention. Further, they did not address the various private security methods of crime prevention among upper-middle-class and upper-class Nigerians. This chapter represents an original study of crime prevention strategies in Nigeria.

DEVELOPMENT OF THIS CHAPTER

The data for this chapter were obtained through ethnographic observations by one who was born and raised in Nigeria. Also, many years of criminological study in Nigeria provided participant observation data to describe crime control and crime prevention strategies in the country. Two phases of data are presented in this chapter—traditional methods of crime prevention and formalized modern methods. Under the formalized modern system of crime prevention are the colonial and postcolonial dimensions. The traditional system of crime prevention through environmental design pervades both phases.

CRIME PREVENTION IN PRECOLONIAL NIGERIA

The year 1849 is generally regarded as the end of the precolonial period in Nigeria. The whole of Nigeria at this point was *gemeinschaft* in form. That is, the society was loosely organized through kinship and common traditions. It was not yet conceived as a country or a nation. In northern Nigeria, the communities were ruled by emirates (Islamic kingdoms) under an emir (Nigerian Islamic king). In western Nigeria, the communities were organized in *Oba* kingdoms. In eastern Nigeria, each community was autonomous. There was a total absence of chiefs, *Obas*, or emirs, to rule the communities among the Igbos of eastern Nigeria, as one would find in other parts of Nigeria. The Igbos believed that there was a "king in every man" (Henderson, 1972). In effect, the head of every household among the Igbos was subject to no one and owed allegiance to no chief or other authority.

Personal Security

In all parts of precolonial Nigeria, the head of every household provided physical security for all its members through the architectural design of the residence and the construction of wooden fences and masonry barriers (walls of mud, bricks, or stones). The fences or masonry walls were circular or rectangular and surrounded all the residential and nonresidential buildings in the compound. There was usually only one door or gate for both exit and entrance to the compound. After the adults left the compound to do farm work in the morning, this security door would be locked. Baby sitters who were usually inside the complex were warned not to unlock the gate or door for any stranger. This front door or gate of the compound's security wall was also locked at sunset every evening and remained so all night long. No one was expected to visit at night, except if there was an emergency in the neighborhood. The compound's security wall sometimes had a short rear door, but this door was always locked both day and night. The rear door was designed to provide access to the gardens at the back of the compound, and it was built short and narrow to make it difficult for a burglar to escape easily, with or without large items belonging to the owners. In most cases, compound walls made of wooden or bamboo fence barriers had no rear entrance-exit gate.

Access control was the common form of environmental design in precolonial Nigeria. Heads of households allowed only one major entrance to their compound security walls or fences. If any footpath was created by members of a household, outsiders were not allowed to use it to come toward the compound. Most of the time, a cross bar was placed across the footpath to warn outsiders not to use it. Some heads of households would also place a leashed dog tied to a small tree adjacent to the footpath to further discourage outsiders from approaching. The idea of employing security guards was not yet conceived, because Nigeria was then a simple, stateless society. Only emirs or *Obas* had a staff of servants who served as vigilantes to protect their kingdoms.

Security of Private Property

The head of a household could use his presence to warn potential night burglars not to target his domain by letting out a series of high-pitched, vocal, exclamatory sounds while either standing on the ground or climbing a palm tree at night. The exclamatory sounds could be sounds of joy or of great excitement or it could be a war song. Moreover, as most burglaries in Nigeria are carried out at night, he could shoot his gun in the air as a warning. Other precautionary tactics were used too. When the head of a household suspected chances of night burglary in the village, he could beat his ancestral drum pretending to appease the spirits of his ancestors. By doing so, he gave notice to the burglars in the vicinity that he was around. It was (and still is) a cultural truism in Nigeria that as long as potential burglars in the town or village knew that the head of a household was at home, they avoided his compound.

Some efforts to prevent criminal victimization in Nigeria involve the para-spiritual realm. For example, property could be saved if it were placed in a Shrine Grove (the altar of the town or village). The Shrine Grove is the home of the community's god—the god of their ancestors who had existed and was worshipped from generation to generation starting beyond the reach of living memory. Some people (Igbos, Yorubas, and others) believe that if anyone steals something saved in a Shrine Grove, that person will die a sudden death or face a mysterious illness leading to the person's death and subsequent burial in a tabooed (evil) forest.[1] There was no exception to the type of property that could be saved in a Shrine Grove.

Security of Real Estate, Cash Crops, and Fruits

Real estate, cash crops, fruit trees, and movable possessions outside the home were secured from vandalism and criminal intervention by using the leaves of sacred trees such as *uboldia* (*umune*) or the youngest leaves of a palm-frond (*omu nkwu*). Every native of the locality would have known the evil consequences of tampering with any property that had any of those sacred leaves placed on it. The negative consequences were the same as those of stealing properties saved in a Shrine Grove. In fact, anything displayed for sale could be secured from theft with any of the sacred leaves. Additionally, some people used ash from their kitchen to secure a property left on a roadside or displayed for sale by an absent seller. The belief was that stealing anything that had kitchen ash on it would render the thief to ashes as soon as the thief ate it—whether cooked at home or elsewhere. That meant that the person would die suddenly and go back to the ashes from which he was created.

Village Security

In precolonial Nigeria, the community or village council was the highest court of justice. Also, the courts of the emirs and *Obas* were the highest in their kingdoms. Each village was made up of people who were united by ties of consanguinity. A town council, as we know it today in Nigeria, was a rare or nonexistent entity. The village or community council was headed by the oldest inhabitant or a person who had inherited the leadership from his father. Each village constructed various barriers against outside invasion or invasion by burglars from neighboring villages. The village council constructed physical barriers for security at various pedestrian access routes to the village, especially those leading to unfriendly villages. Routes to friendly villages were kept open. There were no wide roads because there were no automobiles. Village councils established rotational community guards who stood on alert on strategic routes to the village, ready to apprehend any strangers or burglars entering at night. The use of sentries was a frequent practice whenever waves of burglary and robbery took place.

During the daytime, every adult male was expected to watch for strangers and especially for any notorious criminal wandering around the village, particularly around the market square. It was the job of the vigilantes to find out why the stranger or ex-offender was in the village. If they saw such a person shopping around in the village market, they watched carefully until he left the village. Anybody caught offending in the village was killed without trial and buried that night.

It was (and still is) a common practice that very well-known burglars from other villages or towns were confronted and warned not to come to the village again. If someone violated this warning, the person was beaten up by a group of village vigilantes. There was no freedom of movement beyond one's own village. That was the nature and form of policing in a stateless society void of a formal police agency, modern judiciary, and prison system.

CRIME PREVENTION IN COLONIAL NIGERIA

In 1849, when Lagos was annexed and subsequently colonized by the British, the formal system of policing began. The Lagos Constabulary was established to control and prevent crime at the colonial trading posts, residences, and the colonially controlled urban centers. When the British government granted a charter to the Royal Niger Company Limited in 1861, a paramilitary police force was established to ensure the security of the colonial territory, to prevent crimes, and to keep any resistance from various kingdoms, emirates, and towns under control. The paramilitary police force of the Royal Niger Company stationed companies and squads of the force only at urban areas and seaports. Over 95 percent of Nigeria, which was rural at the time, had no colonial constabulary controlling crime. Consequently, the precolonial systems of crime prevention and control were still in vogue.

The Royal Niger Company's charter was revoked by the British government in 1899, and on January 1, 1900, the British government took direct administration of Nigeria. In that year, the British colonial governor established the modern police system. After the amalgamation of Northern and Southern Nigerian Protectorates in 1914, the modern Nigerian Police Force (NPF) was established. It was at this time that British colonial administrators introduced the provincial and district political divisions. Squads of the Nigerian police force were stationed at every provincial and district headquarters, which were the residential quarters of the chief administrative colonial officers. Although the colonial police were located at the district headquarters from 1920 to 1960, the traditional systems of crime prevention and control were retained in most areas of Nigeria.

The colonial administration focused on British economic interests. In effect, at the time all crime prevention strategies concentrated on the security of life and property of the colonial masters. The rural areas were ignored. Thus, the traditional system of policing and crime prevention predominated. In the absence of colonial rural policing, heads of households and the village councils continued to use the traditional methods of crime prevention and control. The village councils and their subjects saw the colonial police as an agency of political repression and economic exploitation, not one of crime control and prevention.

Throughout the period 1950–1960 when Nigeria received her independence from Britain, the Nigerian Police Force engaged in various crime prevention activities, such as patrolling the highways, streets, harbors, and shopping centers. During the 1950s the Nigerian regional governments were granted self-government. The Nigerians who were the Premiers of their regional governments were able to establish regional police units. The Okpara Mobile Police Force was established to prevent international smuggling through the then French Cameroon and the Spanish island of Fernando Po (now Equatorial Guinea) and to stop highway robbery.

During the period of semiregional autonomy (1955–1959), up to 90 percent of the Nigerian society was rural and the traditional system of crime prevention still predominated. The police were located very far away from the rural towns and villages. When a serious crime, such as murder or aggravated assault, was committed, the village head (a chief created by colonial masters), the victim, or relatives of the victim would invite the police to the rural village to make an arrest.

In urban areas, residential homes of traders were secured with masonry barriers of bricks, cement blocks, or stone walls. The residences of very rich traders and wealthy traditional rulers were used by the colonial administration for the British system of indirect rule.

The colonial administrators and European traders protected themselves from burglary and other types of criminal victimization in three ways: residential segregation, barbed-wire fences surrounding their residential quarters, and the employment of security guards. All European business centers were protected from crime by use of security guards. Additionally, the colonial government posted squads of police at the entrances to the European-owned companies. The iron gates at the front of the residences of the senior colonial administrators, such as the Governor General, the Regional Governor, the Provincial Resident, and the District Officer (DO), were protected by high, painted cement walls and armed guards.

CRIME PREVENTION IN POSTCOLONIAL NIGERIA

The postcolonial period in Nigeria began on October 2, 1960. After Nigerian independence was established on October 1, 1960, the police force came under the control of Nigerians. Within three years, the strength of the Nigerian Police increased somewhat higher than its level during the colonial era. However, the number was still not sufficient to police the rural areas adequately. The division of the former colonial districts into additional jurisdictions brought the police closer to the rural towns and villages than ever before.

In general the crime prevention activities of the Nigerian police concentrate on sporadic patrol of the highways and on the mounting of roadblocks to check for stolen vehicles, automobile and driver's licenses, and traffic regulations. The frequent patrol of the neighborhoods to establish police presence in the cities and to catch criminal offenders in the act that are common in western urban centers is not characteristic of Nigerian law enforcement strategy. Instead, the Nigerian police resort to static law enforcement in which the police do not patrol the neighborhoods looking for legal violators. Usually, the Nigerian police simply stay at the police station waiting for victims or concerned citizens to report crimes (Ebbe, 1989, 1996).

In 1996, the federal government established a new crime prevention unit called Operation Sweep. This unit consists of a paramilitary force created out of the Nigerian Police Force that is given military training and charged with the responsibility of patrolling all the major highways, busy seaports, and large city neighborhoods notorious for harboring dangerous criminals. The Operation Sweep unit is designed to make its presence known and to deter criminals from victimizing businesses and individuals.

Personal Security (1960–1998)

Methods of ensuring personal safety have changed in Nigeria since 1960 largely because people have more money. Nigerian independence created many job opportunities when the colonial administrators and the British civil servants were no longer at their posts. Moreover, international trade between Nigeria and other countries grew by leaps and bounds. In effect, there was a tremendous increase in the standard of living. Wealth started to trickle down even to the rural villages. Heads of households in the villages, who are subsistence farmers, began to construct better masonry barriers of bricks and cement blocks for crime prevention. Pieces of broken bottles were glued on top of cement block walls surrounding a nuclear family, a polygamous family, or an extended family, to discourage criminals from going over. Furthermore, at this time, the traditional 4-foot high mud or clay antiburglary barrier walls gave way to 6- or 7-foot cement block walls, again topped with pieces of broken bottles.

In the 1970s, following the oil boom, upper-middle-class and upper-class Nigerians introduced innovations in private security. The masonry barriers of their residential complexes began to resemble splendid maximum security prisons. Their block walls were capped with barbed wire rising 2- to 3-feet high and running throughout the wall. Some, instead of using barbed wire, began to top the rectangular walls with artistically designed fences. To this day, some of these compound walls are so high that nobody can see the roofs of the houses behind them. Also, unlike middle-class barriers, the upper-middle-class and upper-class cement walls have 12- or 18-inch thick blocks, instead of the regular 9-inch blocks. Additionally, upper-middle-class and upper-class Nigerians employ armed personal guards to watch their homes day and night. To further ensure freedom from crime, in a country fraught with hired assassins, upper-middle-class and upper-class Nigerians hire chauffeurs and bodyguards to accompany them wherever they go.

Local Government Areas

A *local government area* in Nigeria today is equivalent to a county in the United States. Each local government area has a police subheadquarters. The LGA Council uses the police to control and prevent crime in its jurisdiction. The towns and villages under an LGA can go to the LGA police depot and report crime or invite the police to investigate a criminal incident. In each LGA, the police mount roadblocks at major roads leading to the LGA to check the identity of persons driving into or through the area and to inquire about a stranger's mission. Many armed robbers from different local government areas have been apprehended by these roadblocks.

The LGA Council encourages towns and villages under its jurisdiction to continue to use the informal, traditional methods of crime prevention to help the police. Also, the LGA Council unites the police, rural town councils, and village councils for successful crime prevention strategies and offender apprehension endeavors.

Village or Town Council

At the village or town council level, the traditional methods of crime prevention are still in vogue. Wooden and bamboo fences are usually found at poor people's (heads of households) compounds. Also, the construction of mud and clay walls as antiburglary barriers at lower-class compounds are common throughout all the regions of Nigeria. As of May 1998, over 65 percent of all Nigerians still live in the rural areas. Nevertheless, in the midst of the wooden fences and mud walls, one can find magnificent upper-middle-class or upper-class residential complexes. Any Nigerian with wealth who lives in a city but has not built a residential house in his native village is regarded as "a failure." This is why Nigerian cities become almost empty during traditional festivities. Everyone living in the city goes to the village of her birth, where her kinsfolk still live. There is no residential segregation or "restrictive covenant" in real estate for the nobles.

Youth Associations

In every town or village in southern Nigeria, there are progressive youth organizations. Every young, adult male, 18 to 35 years old, aspires to belong to one of them. These youth organizations prohibit their members from involvement in criminal behaviors. Any member who is found guilty of criminal behavior of any magnitude is excommunicated. Expulsion from a youth organization is the worst thing that could happen to a young man in Nigeria. Young men who are not members of any youth association are regarded as outlaws, subordinates, and social misfits. Consequently, youth associations operate as crime prevention institutions, because of the deterrent effects they have on their members, neophytes, and prospective members: No person with a criminal record is ever admitted to any of the youth organizations.

Business and Industrial Security

Since the oil boom of the 1970s, there has been a tremendous increase in business activity, including manufacturing. When the oil boom collapsed in the 1980s, a large population of unemployed youths resorted to burglary and highway robbery. Private businesses, corporate manufacturing companies, and statutory corporations became prime targets. Consequently, all manufacturing complexes today are fenced with barbed wire and secured by platoons of armed guards, both day and night. Some companies use welded 1-inch iron rods to barricade the entrance to their main offices.

Before entering a Nigerian bank, one must pass through a guarded iron gate located roughly 50 feet away from the main door. In every Nigerian bank, whether rural or urban, armed guards stand at the gate and the entrance door to the bank's main office. For large banks, there is always a squad of four or five armed guards watching everyone who comes in and goes out of the bank. Consequently, no Nigerian bank has reported any incident of bank robbery in over 20 years. Most banks in Nigeria have electronic devices that keep the banks safe from night burglary. Some electronic devices are designed either to hold the burglars from escape after entry or to kill them instantly.

SUMMARY

Crime prevention in Nigeria involves three phases of crime prevention (Lab, 1997). *Primary prevention* involves active social organizations, environmental design of homes, neighborhood watch units, punishment for general deterrence, private security, and public education. *Secondary prevention* emphasizes identification of people in the local government areas

(LGAs), reports on strangers found in the villages, identification of known burglars in the villages and urban areas, and community policing. Targeting high crime areas of the cities and using an Operation Sweep unit to go after the notorious criminals also appear here. Finally, *tertiary prevention* involves incapacitating the offender by excommunication, expulsion from the neighborhood, and public ridicule, or "shaming the offender" (which is effective in both specific and general deterrence).

Every society uses different types of punishment to deter potential offenders. Deterrence, in effect, means crime prevention. The effectiveness of any form of punishment, however, depends on two factors: the certainty and the swiftness of its application. In this vein, Nigeria exacts the death penalty for murder, armed robbery, counterfeiting currency, night burglary, and treason. The government elevated these crimes to capital offenses because it is believed that the fear of death by a firing squad serves as an efficient crime prevention mechanism. The effectiveness of the death penalty in Nigeria is ensured by certainty of application and swift application to convicts. Over 80 percent of Nigerian nationals support the death penalty.

Nigerians, like people of most other African nations, believe that the best place to treat a criminal offender is in the community, not prison. The colonial administration, however, introduced the prison system in Nigeria, and the postcolonial governments still use it to discourage offenders. In Nigeria, the notion that a person is in prison disturbs the collective conscience of the offender's community. The offender's relatives feel ashamed that one of their own is in prison. In effect, communities in Nigeria, as elsewhere, try to keep their members from becoming involved in criminal behavior. At the same time, individuals in rural towns and villages try not to bring shame and disgrace to their communities by avoiding getting involved in criminal behavior and being incarcerated. This fear of embarassment over imprisonment for criminal activity becomes a very effective crime prevention mechanism.

In conclusion, crime prevention strategies in Nigeria are evolutionary. As criminals continue to specialize and advance their modus operandi, the Nigerian government and the masses also work to design new methods to contain them.

ENDNOTE

1. The people of southern Nigeria (like the Igbos and Yorubas) bury the dead on their own property. For a deceased person to be buried in an evil or tabooed forest is a curse. Such a person is not expected to reincarnate—the person's cycle of life on earth and in the land of the spirit world is ended.

SECTION 3

The following three chapters deal with community policing and crime prevention in emerging democracies. Each of the countries is a former member of the Soviet Union, and each is facing the demands and trials that come with major political and social changes. Primary among these challenges are political corruption and civil war. The problems facing the police and criminal justice systems are very different from those in other countries.

In the past, social control was the province of strong centralized authorities tied closely to the political regime, and often involving military forces. The police and criminal justice systems often focused on the needs of the state, rather than the needs of individuals. A key concern was the stability of the communist governments. Thus, the suppression of dissent was a major task for the police. The fall of the Soviet Union and the attempt to move toward a more democratic form of government necessarily shifts the role of social control agents. Unfortunately, much of the populace still retains its fear and distrust of the police and the courts, making it difficult to develop initiatives that require close interaction with the citizenry. The police and the authorities in these countries recognize the potential for community policing and crime prevention, but the establishment of these approaches is problematic.

Another problem impacting on the police and social control agencies in these countries involves the tough economic conditions they face. The movement to free markets and democracy has introduced a level of economic uncertainty not known in recent decades. Not only do the poor economies contribute to the growth of crime and deviance, they also make it difficult to field a meaningful response to these problems. Many agencies cannot pay their employees on a regular basis, much less fund new community policing and crime prevention initiatives. Nevertheless, the inclusion of the citizenry into policing holds the potential of alleviating some of the financial burden of the social control agencies. This "catch-22" situation is a difficult one to tackle. The chapters that follow offer some insight to that process.

CHAPTER 11

CRIME PREVENTION IN YUGOSLAVIA

Branislav Simonovic
Miroslav Radovanovic

The last decade of the twentieth century was one of the most difficult periods in the history of Yugoslavia. The former Yugoslavia disintegrated into civil war. Although actual combat did not take place on the present-day territory of Yugoslavia, the consequences of war are felt there. The country is still in a deep political crisis characterized by conflicts at various levels and intensities. Further, the economic situation in the country is very tenuous. Large industrial establishments are either idle or work at a reduced level (10 to 30 percent of capacity). Tens of thousands of workers are underemployed. The very existence of the majority of the population is seriously threatened. The situation is aggravated by approximately 600,000 refugees. With the country in transition, the legal system is still not stable. Statutory discrepancies between the Yugoslavian Constitution and the various laws create legal snarls for the police.

The disintegration of the country and the onset of the war brought about a 20 to 30 percent increase in crime. New forms of serious deviant behavior have emerged, and even petty crime has increased. Criminologists claim that conditions conducive to the committing of crime have combined in an unprecedented manner (Ignjatovic, 1995). Yugoslavia, at this moment, is an example of anomie in action.

HISTORY OF CRIME PREVENTION IN YUGOSLAVIA

The development of crime prevention in the former Yugoslavia began more than forty years ago. In the early 1960s, Yugoslav criminologists began to advocate that the fight against crime had to have a broad base and that the application of repression alone could not produce satisfactory results.

> We have come to believe that the focus of the fight against crime should be based on social prevention consisting of various social activities and the activities of the state and social organs and the citizens and directed primarily at the detection of direct objective and subjective causes of criminal actions and other phenomena of social pathology. What we are talking about is the application of various social, economic, educational, cultural, health and other similar measures aimed at the removal of direct sources of criminal actions (Milutinovic, 1964:26).

The fight against crime cannot succeed unless it is directed against the crime-inducing conditions and causes. Preventive measures must focus on cutting off the social roots of crime (Jasovic, 1966). Society itself is, to a greater or lesser degree, responsible for every criminal action (Jeremic, 1963). Thus, prevention is an expression of a deeply rooted humane attitude toward people and a reflection of the need for broad social action (Krivokapic and Boskovic, 1984).

Yugoslav criminologists believe that preventive and repressive measures are two necessary, complementary, and functionally interrelated areas of criminal policy. The domination of one over the other should not be allowed. Social prevention by itself, without the threat of punishment, would be an absurd way to guard society, as would be the insistence on the penal function alone (Milutinovic, 1976).

In Yugoslav policing science, prevention is divided into social and criminal prevention. Theoreticians back the view that the criminal organs (courts, prosecution, and the police) should not undertake repressive measures in the absence of preventive ones. Yugoslav criminal law proscribes that the purpose of punishment at the level of individual prevention is reeducation and rehabilitation of the perpetrator. On the general prevention level, the purpose of punishment is to serve as a deterrent to others, to strengthen ethics, and to encourage social responsibility and discipline among citizens. To put these goals into effect, however, is not an easy task.

Crime prevention in the former Yugoslavia was part of the concept of social self-protection designed to integrate protective functions into the society. Established in the 1950s, it was within this system that the necessity of prevention from antisocial and asocial phenomena was developed. It was claimed that this concept was considerably broader, more meaningful, and more flexible than any other system of protection. It was intended to protect society not only from criminal behavior, which is the province of criminal policy, but also from all other types of deviation and antisocial behavior (Milutinovic, 1976).

The concept of social self-protection arose from the concept of self-management, by which the owners of the means of production were the workers employed in production. Basically, it was protection from self-management socialism and the means of production. It enjoyed positive support from the media and those in positions of authority. The intent of its creators was that it be applied primarily in work organizations and in the local communities.

Although a great deal of effort was invested in this concept, it failed to show any protection of socially owned property because, among other things, the authority of the self-management worker, established in various industrial enterprises, was limited to the physical protection of the *means* of production. Also, it failed completely to protect socially owned property from economic and financial crime. When crime was committed, protection was carried out by the traditional repressive organs of the state. Strategically, prevention was left to the workers who were also the owners. Since this did not function, there was no satisfactory preventive activity. In practice, a pronounced imbalance existed between repression and prevention—in favor of the former.

A number of criminological research institutes were established in the early 1960s. They carried out projects on the causes and prevention of various types of crime. Unfortunately, there was no coordination among the research activities, their results, and practical applications to criminal policy. But, despite these shortcomings, some crime prevention activities in the former Yugoslavia were carried out.

In Yugoslav criminology, prevention is divided into a number of different approaches, each of which emphasizes a different aspect of the field. Attacking the sources of crime is divided into (1) general social prevention, (2) specific social prevention, and (3) individual prevention (Milutinovic, 1984). The first comprises broad social activities intended to improve the social environment and make living conditions more humane. Here, the crime-inducing causes and conditions are restrained. Direct actions vary greatly, from employment programs and lessening of economic hardships to cultural and educational activities directed toward the young (for example, organized after-school sports and other activities). The application of most of these measures requires a relatively strong economy: They require long-term investments and their positive effects take time to be felt. Because they create a broad, positive basis for a healthy society, they should be the primary goals of any well-conceived crime prevention policy. Such crime prevention measures persisted until the disintegration of the country. Today, very little is being done in this area.

Specific social prevention attempts to remove the crime-inducing factors that exist in a specific social environment. These measures are directed at causes and conditions affecting certain categories of persons. Specific social prevention is most successfully applied at the local level. In Yugoslavia, this type of preventive measure is aimed primarily at minors (for instance, by means of school lectures on the dangers of drug addiction or on traffic safety, organized day care for school children, asylums for orphans, and financial help for economically threatened families).

Individual preventive measures are intended to remove crime-inducing situations from specific persons. They could be applied to persons who have not as yet committed a crime, but show deviant behavior (for example, truancy, aggression, or petty theft) or to persons living in unfavorable conditions (for example, abandoned children). The application of this type of prevention has been sporadic and insufficient. However, most authorities agree that these steps should be undertaken with children and minors, and they are applied as educational and humanitarian measures. Individual preventive measures can also be directed at persons who have already committed a criminal act and now are serving, or have served, a sentence. Such measures should facilitate rehabilitation and inclusion into society, by means of employment programs for ex-convicts and work camps for juvenile delinquents. In spite of a number of problems, especially financial, these programs have produced positive results.

Attempts to curb juvenile delinquency can be divided into (1) programs for modifying and humanizing the social environment, and (2) programs directed at minors who show signs of maladjustment or have committed a criminal act. These steps are designed to produce psychological and social changes that would make the youngsters resistant to delinquent behavior. In Yugoslavia, many more programs are oriented toward the resocialization and rehabilitation of a maladjusted minor than there are programs oriented toward the improvement and humanization of the social environment. Particularly lacking is work with the family of the minor, which as a rule is the source of the child's problems (for example, separated parents, criminals, drug addicts, or alcoholics within the family circle). On the other hand, there are no coordinated actions between economic and industrial complexes, whereas there are many social problems that cause an increase in delinquency (Milutinovic, 1965). There have been positive results with only a small number of delinquents.

With the disintegration of the former Yugoslavia, the concept of social self-protection was ultimately abandoned. The problem is that no other general concept has as yet been established and that the state relies primarily on repression.

PRESENT-DAY CRIME PREVENTION PROGRAMS

Intervention in Crisis Situations

Recently, because of a large increase in juvenile delinquency in Yugoslavia, a coordinating body was established under the auspices of the Assembly of Belgrade to deal with the problem. This body consists of representatives of the judiciary, police, schools, health institutions, and experts from the Belgrade Center for Social Work. In July 1995, this coordinating body launched a program for the prevention of crime committed by criminally irresponsible children, entitled Intervention in Crisis Situations. This program is intended to deal with first-time offenders under the age of 14 who are not subject to the provisions of the criminal law. The program envisions a close coordination between the Section for Prevention of Juvenile Crime from the Department for Juvenile Crime of the Internal Affairs Service of Belgrade and the Belgrade Center for Social Work. It is a team effort consisting of an inspector who supervises activities aimed at the prevention of juvenile delinquency. The inspector is assisted by two social workers. A member of the team is constantly on duty at the offices of the Section for Juvenile Delinquency.

The program consists of several phases. Phase One involves the detection of the offender. After detection, contact is established with the team, and the offender is brought in for further treatment. Phase Two is the establishment of contact with the family. Following an information-gathering interview with the child, the team immediately visits the family to inform the parents (guardians) of the criminal act committed by their child, to explain the reasons for their visit, and to establish trust and cooperation for further action. If the parents agree to cooperate, the team places the family into an experimental or control group (depending on the time order of reports reaching the team). The experimental group family is offered help by the Marriage and Family Counseling Service. If the family accepts the offered help, a date for their visit to the counseling service is set. The control group family initially receives no help. After a certain time period, however, they will be called to report to the Center for Social Work in their community where a team of experts will meet with them and try to help them solve their problems.

In Phase Three, therapeutic work with the experimental families is undertaken. In five to six sessions of family therapy, experts from the center help the family overcome the crisis brought about by the criminal act and attempt to establish better-functioning patterns of interrelationships. Phase Four involves monitoring recidivism. Recidivism is monitored in order to evaluate the success of the program and give help to the child. So far, recidivism of only 10 percent has been recorded. Phase Five consists of a follow-up after the initial 3-month period. The team visits the family in order to provide a professional assessment of possible changes and to gather information for the program. Finally, Phases Six and Seven consist of an analysis of the program's results to see how successful it has been and to decide which aspects of the program could be applied to other families.

Day Care for Minors

A number of schools in the environs of Belgrade have organized whole-day sojourns for minors exhibiting excessive predelinquent and criminal behavior. Pedagogues and psychologists from the Center for Social Work organize the activities of the pupils, help them do their homework, and supervise sports and recreation activities. The program helps at-risk children with their school duties, with the planning of their leisure time, and with their socialization.

"School"—A Program for At-Risk Children

Since 1993, in Belgrade and other cities in Serbia, a program entitled "School" has been in operation. It is carried out by the police in cooperation with educational institutions. Its objective is to prevent or curb all asocial behavior in the schools and their neighborhoods. When there is any kind of conflict in the schools (for example, individual or gang fights, thefts, or any other disturbance of the instructional process), the police are called in. A police inspector takes part in an information-gathering interview with the offending pupils' parents. Increased attention by school officials is directed toward pupils showing predelinquent behavior. Also, the schools are charged with the obligation of carrying out preventive activities and reducing juvenile delinquency. The police participate through tasks such as organizing lectures on the dangers of drug use, increasing the frequency of patrols (both uniformed and undercover) in the vicinity of the schools, and identifying suspicious persons. It is believed that this program helps reduce aggressive behavior in and around the schools and decreases the number of thefts within schools. One problem is that school officials may cover up the criminal behavior of their pupils and attempt to solve the problems by themselves in order to preserve the school's reputation. Sometimes, they fail in their attempts and ask for help only when problems have exacerbated.

"Humane Patrols"

In Belgrade, from time to time, "humane patrols" are carried out. This action deals with Romany children and vagrants who engage in begging. For example, during 1993, many Romany children from Bosnia were brought by their relatives and abandoned in the streets of Belgrade to beg. Even today, any children caught begging are taken to orphanages for children and adolescents. Their parents are sentenced, usually, to 30 days of imprisonment.

Associations for Crime Prevention

In some cities of Serbia, voluntary citizen associations were recently organized to help with the prevention of crime and other social evils afflicting the young. In Belgrade, an association seeks to protect children and women, and another to provide aid for victims. However, these associations are seriously underfunded. Except for their good will, they have little else with which to work. Their cooperation with the state (for example, the police) is also questionable.

Police Prevention

The police implement, as part of their public safety activities, a series of measures with a pronounced repressive and preventive nature. Their activities are divided into: (1) direct prevention (a collection of all the measures undertaken by the police within the scope of their activities) and (2) indirect preventive activities (aimed at prevention involving outside factors and initiated, directed, and coordinated by the Internal Affairs Service).

Police control is directed at crime-inducing areas, such as the premises and meeting places of criminals, because these are the places where criminal acts could be planned or committed or where wanted criminals can be found. Control also is directed against persons generally believed to be disposed toward criminal activity (like recidivists, registered criminals, vagrants, prostitutes, and inmates on leave). Such people are subjected to more intensive supervision and police monitoring (special observation). However, many problems accompany these measures. For instance, poor cooperation by the general public and other institutions and not enough police in the area result in higher recidivism (Boskovic, 1994; Krivokapic, 1996).

During the mid-1990s, preventive police activity had been neglected. There were almost no foot patrols and neighborhood police activities. This had an adverse effect on police-citizen cooperation. The absence of uniformed officers meant that the police could not get to know the local area, to see how safe it was and to establish cooperation with the local citizenry. The consequences of this unfortunate police practice resulted in alienation of the police from the citizens, inefficient supervision of criminals, and increased numbers of unsolved crimes. Only in the late 1990s was policing of areas reintroduced.

The New Instructions on police work insist on knowledge of the local safety situation, measures for increased supervision, neighborhood policing, patrol activity, and traffic regulation (Skakavac, 1997). Even in light of these new instructions, the role of the police at the local level is mainly state-oriented. No cooperation between the police and the citizens and their associations has occurred, which is already manifesting itself in increased passivity and restraint on the part of the police.

In addition to political and organizational reasons for the lack of cooperation at the local level, the literature advocates a rather narrow view of police prevention that is not to be identified with the removal of crime-inducing causes (that is, by social prevention of crime). The means employed would involve routine police work, such as observation, investigation, and supervision services (Vodinelic, 1985). The police should be active even in areas in which crime has not yet occurred but which could be conducive to crime. This is where preventive measures could play a vital role. This activity, by its nature, implies taking steps outside traditional police control and investigation, such as educating potential victims, providing counsel to citizens, and working with at-risk children and adolescents.

Periodically, the police in Yugoslavia undertake a series of repressive-preventive actions, such as increased action to prevent automobile thefts and detection of the perpetrators, and legalizing and confiscating weapons. The problem, however, is that the emphasis is always placed on repression, while prevention is mentioned only in passing and has a second-rate role. The tasks, the organization of the necessary activities, and the personnel to carry out preventive or repressive activities have not been delineated. Because of this, prevention is neglected. That is, prevention is carried out only through repression as its by-product, and not separately in its own right.

CHALLENGES OF IMPLEMENTING CRIME PREVENTION

For several reasons crime prevention has not made much headway in Yugoslavia. Some of the most important factors are discussed below.

Legal Problems

The constitution of Yugoslavia from 1963 to 1974 charged the courts with monitoring and studying social relations and conditions relevant to their functions, and with offering suggestions for the prevention of phenomena that might be harmful to society. The courts were also obliged to inform the assemblies of sociopolitical communities about problems with the implementation of the law, as well as about conditions that might contribute to crime. This constitutional obligation was also prescribed by the Law on Courts.

A number of meetings of judiciary representatives were held to discuss the implementation of these provisions. It was decided that the regular courts should undertake compliance with the Constitution. It was pointed out, however, that there were practical problems affecting the implementation and success of these activities. Primarily, the sociopolitical communities maintained a passive, if not indifferent attitude toward any interventions, analyses, and information supplied by the courts and the public prosecutor's office. It was felt that, in practice, there was no adequate feedback to the courts on what had been done to remove crime-inducing causes and conditions. Although the courts had repeatedly called attention to particular weaknesses, those weaknesses were not addressed, and criminal behavior persisted. There were egregious examples of negligence. For instance, after a number of accidents at an unprotected railroad crossing, the courts finally pointed out the danger. The incompetent authorities, however, did nothing to secure the crossing. Consequently, a train collided with a bus full of passengers, with a large number of victims (Loncarevic, 1987).

In the current provisions of the constitutions of the *Federal Republic of Yugoslavia* and the *Republic of Serbia*, and in the provisions of the existing Law on the Courts, the word "prevention" is not even mentioned. Obviously, the legislation has been adapted to current practice, instead of the other way around. What has been done with the legislation is a step backward. Previous solutions were basically sound, but they lacked legislative support and the legal mechanisms for implementation. It would have been better to provide a legal basis for victims or public prosecutors to initiate a court action for compensation of damages against the authority that had not acted upon warnings issued by the court. The repressive function of the police has been precisely defined by various statutes, regulations, and instructions, but prevention measures have been mentioned only in passing. It is felt that no obligation exists to undertake prevention (Krivokapic, 1986).

Popular Attitudes Toward Prevention and the Police

It is widely held that the attitude of citizens toward crime prevention in Yugoslavia has never been encouraging. The average citizen is not interested in crime and its prevention until he himself is involved. Unfortunately, "the safety awareness" of the citizens is not sufficiently developed. The media have done almost nothing about it. Very few substantive TV programs approach the problem and discuss prevention measures. Recently, however, a sudden increase in the number of criminal acts, with the onset of the war, has made citizens aware of the severity of the problem and the potential danger to themselves. At the same time, two new problems have arisen. They can be summed up as a lack of specific prevention programs and a lack of cooperation and trust between the police and the citizens.

What little respect the police enjoyed in the past has been seriously threatened by, among other things, several instances of the use of massive police action against citizens who, expressing their affiliation with the opposition, have demonstrated against the government. Such standoffs have been recorded by several Yugoslav police units and legal experts and have resulted in pronounced animosity. For example, Krstic (1997:208) states that "there is no doubt that the police [are] not liked by the citizens. It is an incontrovertible fact that, in recent years, a gap has been created in the trust between the police and certain strata of society." And Pihlen (1997: 388–389) offers the following analysis:

> Our police, in recent years, have lost what little respect they had enjoyed earlier. Citizens have never had any real trust in the police, while today the police are seen as [a force] one should avoid rather than cooperate with and expect anything from, a force that has been turned against them and not toward their interests, individual and social needs. Ever more, the police are identified with the actual, illegitimate regime in power and with some unidentified forms of pseudo-power. Skepticism toward their work and efficiency in crime prevention is particularly emphasized. Because of that, the basic preventive nature of the police function has been suppressed by the other one, the repressive nature.

These opinions point to the consequences, not the causes, of the present-day role and position of the police in Yugoslavia. One cause of low esteem among the police is the abuse of police function. That is, the police have often been "used" for purely political goals. A truly democratic society and a police force with unquestionable integrity are the *sine qua non* of successful preventive activity. However, the police themselves are often not above reproach. They are sometimes criticized for arrogance in their dealings with people, for distancing themselves from the citizens, and for maintaining secrecy as a means to cover up their negligence and lack of expertise (Krstic, 1997). It is generally believed that the police lack what it takes to establish and maintain good contact with the public.

All these factors influence the reputation of the police and their relationship with the citizens. When cases of corruption of police officers are added (Krstic, 1997), one gets a real and fairly dismal picture of the state of the police in Yugoslavia. Citizens are unwilling to give information to the police or to appear in courts as witnesses because they are afraid of revenge by criminals and do not believe that the state can, or is willing to, protect them.

The key question to be asked is how can these circumstances influence crime prevention in practice? One fact is obvious: The relationship between the police and the citizens, as well as their mutual cooperation in the reduction and prevention of crime, are severely disrupted. Those engaged in everyday police activity observe that most citizens have stopped completely cooperating with the police. Today, the police can count only on the help of the victims themselves.

Once again, what should be done to overcome the existing situation? The police and the authorities need data on the public's opinion of the police in order to correct, program, and harmonize police attitudes toward the people, raise the level and quality of their professional activity, and enhance their reputation. The public, too, must be informed about what the police are trying to accomplish and the problems they face. Citizens have the right to control the police, since they are the ones who provide finances. The realization of such a relationship requires, first of all, the establishment of a truly democratic society, a removal of all political influence on the police, and a change in the concept of police activity, primarily at the level of a local community. The experience of other nations could be used as a model.

Problems of Organizational Nature

It has been known for a long time that prevention has become an orphan of police service, despite the fact that it is the source of creative work and requires the attention of the best personnel (Vodinelic, 1985). One reason for this is the belief that prevention has limited applicability (this argument is not totally valid since the most crime-inducing behavior can be treated with preventive measures). Another reason is the problem of measuring success. The results of preventive measures, when and if used, are not obvious and attractive in comparison to repressive measures. Therefore, the latter are usually taken as the only measure of success (Krivokapic and Boskovic, 1984; Krivokapic, 1992).

Still another reason is the strictly hierarchical organization of the police, which limits initiative and freedom of action at the local level. The overall structure and mechanism of police activity is focused on repression and a chain-of-command approach. This structure has an adverse effect on preventive activities, self-initiative, and cooperation with local citizens.

The most important reason, however, is the fact that prevention has not been institutionalized and separated from repression. At the level of the Ministry for Internal Affairs, no specific unit is in charge of crime prevention. Within city secretariats for internal affairs, there are no departments for crime prevention. Since preventive activity has not been separated (institutionally and by personnel) from the repressive activity, it is understandable that it is neglected in policing. People forget that it is better to prevent crime than merely to satisfy a quota by solving criminal acts (some of which could have been prevented).

Lack of Good Prevention Programs and Trained Personnel

Lack of research in the area of prevention weakens arguments on the extreme importance of prevention (Krivokapic, 1986). The neglect of prevention proves that the work of criminologists has not been taken into account. Projects of the Institute for Criminological Investigations point in the direction of developing preventive activities. Positive results in the area of police prevention experienced by other nations have also been pushed aside. Similarly, professional education and training programs intended to train personnel for preventive activities are lacking.

Problems Relating to Decreased Economic Power

The great economic crisis that Yugoslavia is experiencing is an additional negative factor that discourages possible crime preventive measures. The biggest problem is the resocialization of offenders. Due to the crisis, it is impossible to carry out good programs of social rehabilitation and reemployment of ex-convicts. After release from prison, these people experience a deep existential crisis with no help from society. Without some assistance they often are forced to revert to criminal activity again.

The treatment of juvenile delinquents is also a major problem. In correctional institutions they go through a program of social and psychological treatment. Their families, on the other hand, due to a lack of funds do not receive any treatment. After a stay in a remedial institution, minors return to the same family which may have turned them into juvenile delinquents in the first place. The prognosis for making good citizens out of such neglected children is poor.

Crises of the Family

Due to the great economic crisis in Yugoslavia, many families are in a very difficult situation. Their outlook for the future is poor. The young are facing particularly difficult challenges. Problems that have been identified are dissatisfaction with both family and school, negative outlook on society, fear for personal safety, and lack of future goals. The use of drugs by minors, even by young children, has increased dramatically. Similarly, a tremendous increase in alcoholism and suicides among the young has been recorded. Aggressive behavior, street fights, and even violent death are on the rise. The number of thefts and dangerous traffic violations have also increased. Several cases of "Serbian roulette"—driving a car at high speed through a red traffic light—have been recorded, often resulting in death.

Because of the lack of funds, centers for social work are not in a position to implement effective programs. In Yugoslavia, the culturally, educationally, and economically deprived population has a need for social education, but the centers for social work are incapable of financing larger programs in this area. Because of strong traditions, families are, as a rule, not welcoming to outsiders. Social workers find it difficult to assist families whose members have shown deviant behavior. As a result, aid to families is either insufficient or nonexistent. Without effective socialization of families, the outlook for crime prevention is not good.

Because of great social, demographic, and economic differences among the people of Yugoslavia, a single program of social prevention is probably not the best solution. Instead, programs designed for specific regions should be applied. The idea of counseling services that will include a large number of families and children also deserves support. Crime prevention through social education must focus primarily on the family.

THE FUTURE OF CRIME PREVENTION IN YUGOSLAVIA

In view of existing criminal policy in Yugoslavia and the great economic and political crisis, we must ask: What next? What is the role and place of crime prevention in the near future? Some authors express great skepticism on the subject.

> The existing social situation, as well as other factors defining the contents of contemporary criminal policy, give rise to the supposition that, today, repressive measures, although of limited scope, are much more efficient and almost the only means of fighting crime. These suppositions, at the time of a

> massive increase of crime, might appear better grounded than the application of preventive measures whose direct results are difficult to prove today (Krivokapic, 1994:116).

This opinion addresses a theoretical dilemma that directly bears on the development of crime prevention. It is tempting to conclude that the only efficient means for fighting crime, in a situation where all sorts of criminal acts are on the increase, is investment in repression. However, such a conclusion is very short-sighted: When criminal acts are on the rise, one must support not only repression, but also prevention. Doing so will eliminate crime-inducing causes and conditions.

The case of Yugoslavia is an example of unsatisfactory results achieved through the application (for a number of years) of police prevention only, and in Yugoslavia this is crime prevention in the narrow sense—prevention through repression. There can be no successful crime prevention without cooperation and coordination of preventive activities between the police and the citizens, and between the police and other social institutions (for example, centers for social work and schools). The development of crime prevention implies certain organizational measures within the state. Prevention cannot be successful unless it is institutionalized and unless clearly defined groups are charged with crime prevention as their basic task.

Two preconditions must be met before prevention can take root in Yugoslavia. First, preventive activity must be institutionalized. For instance, the establishment, within the Ministry for Internal Affairs, of a separate unit that would be engaged solely in the issues of crime prevention is one example. Various professionals would be employed within this unit. Its tasks would be monitoring the spread of crime, designing prevention programs, and guiding the implementation of the programs. In the city secretariats for internal affairs, it would be necessary to establish smaller organizational units charged with local-level crime prevention. Yugoslavia must apply some of the models from countries with already developed crime prevention programs.

The second precondition would be an improvement in the relationship between the police and the public. Serious time and effort should be invested in the creation of a new, improved image of the police that is not linked with the fate of the ruling party and does not make the police appear to be its powerful protectors. The establishment of some kind of external control of the police and programs designed to bring the police close to the people, primarily at the local level, would have a positive influence on public cooperation with the police.

Only with the fulfillment of these preconditions can effective crime prevention be realized in Yugoslavia. The lack of domestic programs and experience could be temporarily overcome by imitating foreign solutions adapted to domestic conditions. Cooperation and integration with international institutions engaged in crime prevention could help in this regard.

Two serious obstacles to the development of crime prevention in Yugoslavia are the lack of the political will to effect it and the persistence of the economic crisis. As yet, the idea of establishing a national institution in charge of crime prevention has not taken root. Perhaps it would be more realistic to insist on the formation of departments within certain ministries that would be charged with prevention issues. The tasks of these crime prevention departments would consist of identifying crime-inducing conditions and causes (so that preventive activity would be directed toward them), designing preventive activities, drawing up concrete prevention plans and programs, providing the personnel to carry out the programs, establishing feedback mechanisms between the departments and local-level programs, and setting up the mechanisms of control. The police would act as expert advisors providing these bodies with data on criminal activity, the various perpetrators and their milieu, the planning and commission of criminal acts, crime-inducing causes and conditions, and the current situation in particular areas. With this information, crime prevention departments within the ministries could move ahead effectively.

Conditions are ripe for the formation of crime prevention departments within the ministries of police, education, finance, legal affairs, public safety, and ecology. A broad basis for crime prevention should be set up, and conditions for unifying these activities should be created within a single institution. Also, a decentralization of the police and a move toward the establishment of a more democratic society should create more favorable conditions for further development of crime prevention at the local level.

CHAPTER 12

CRIME PREVENTION IN HUNGARY: A COMMUNITY POLICING APPROACH

Ferenc Bánfi
Irene Sárközi

On October 18, 1989, the Hungarian Parliament approved a new constitution by an 88 percent majority and thus abolished the People's Republic. The preamble states,

> The Hungarian Republic is an independent, democratic, law-based state in which the values of bourgeois democracy and democratic socialism hold good in equal measure. All power belongs to the people, which they exercise directly and through the elected representatives of popular sovereignty. . . . No party may direct any organs of state.

Ethnic minorities have equal rights and equal educational opportunities in their own language. The single-chamber National Assembly has 386 members made up of 176 elected representatives, 120 allotted by proportional representation, and 90 from a national list. Assembly members serve 4-year terms.

Hungary has one national police force, with a total number of 31,500 officers. In addition, roughly 10,000 civilian employees perform auxiliary functions. In 1997, a government resolution provided for hiring 3,980 more professional staff. It is interesting to note that, immediately before the change of political system in 1989, there were only 20,308 ordinary (that is, not state security) officers and 4,520 civilians within the police (Rendõrségi Évkönyv, 1993:104), the latter also providing services to organs of state security.

The Hungarian National Police (HNP) Headquarters is the focal point of the centralized force, endowed with all the powers needed to direct and control the entire network. A National Commissioner, appointed by the Prime Minister but reporting directly to the Minister of the Interior, is the chief of all police officers. His two deputies, called Directors General, have the responsibility for overseeing the two basic fields of police work—criminal investigation and public security (uniformed). Direction is exercised as a rule (but not exclusively) through county headquarters. There are 19 county police forces in Hungary. The capital force has the same status as county police. Local police stations, usually serving several municipalities, represent the lowest level of the pyramid.

The structural organization of the Hungarian Police is based on the principle of unity. This means that all police officers have the same rights and duties. Of course, this does not exclude the internal division of functions. In practice, officers specialize in criminal or in public security tasks. Those focusing on the latter must contribute to the detection and investigation of crimes by collecting information, securing crime scenes, and performing certain criminal procedure duties.

The Minister of the Interior directs the police. The Minister is responsible for the proper and efficient functioning of law enforcement. The ministry facilitates the carrying out of crime prevention within law enforcement organizations. The competence of the Minister to direct police activities covers any field of policing, including investigations. The limitations on this power are stipulated in the Police Act. For example, the Minister can order an investigation but cannot prohibit the police from launching one. These provisions ensure operational autonomy of the police.

The Police Act provides that police have a general responsibility for crime prevention. The act itself does not give detailed regulation on particular tasks or duties in this field. Apart from crime prevention supervision, no specified institutions serve the preventive function. Actually, crime prevention, as a general duty of the police, was added to the elaborated text of the act only at the end of parliamentary debates.

The Police Act views the police as the dominant actors of crime prevention, as opposed to more modest participation in prevention of administrative offenses. This is a serious mistake because of the social roots of crime. The act does not even refer to the well-known differences among primary, secondary, and tertiary levels of prevention. Primary prophylactics influence general conditions of social coexistence to decrease the danger of crime. Improving the quality of life, for example, certainly contributes to prevention of some types of offenses. Secondary prevention focuses on a more specialized area of criminality (for example, car thefts), and tertiary prevention deals with specific individuals and the dangers associated with certain concrete acts. According to dominant views within criminology, primary prevention does not have anything in common with repressive police activities. In secondary prevention the police may play some role. Tertiary prevention is much closer to the police function, but even here law enforcement must not act alone (Gönczöl, 1996).

One can conclude that the act does not restrict police to work only in the tertiary field. Indeed, some authorities point to an important social policy impact of police decisions. In practice, police run a number of programs, many of them focusing on youth.

CRIME PREVENTION PHILOSOPHY OF THE HNP

Crime prevention is, though seemingly a new idea, far from novel. Ever since crime existed, issues of crime prevention have been dealt with under the specific conditions of the various historical periods. Cesare Beccaria, the outstanding Italian scholar of the age of Enlightenment, can be considered the founder of the scientific foundations of crime prevention. His maxim, "It is better to prevent criminal acts than to punish them," has become a household phrase.

Various crime prevention units and organizations started to emerge in developed western European countries in the 1960s and 1970s. In international organizations (like the United Nations and the Council of Europe), issues of crime prevention surfaced frequently during the 1980s. Of these, the most important was a recommendation adopted by the Ministerial Commission of the Council of Europe in 1987, which stipulates the following for the governments of the member states:

- Establishment of national, regional, or local crime prevention organizations
- The launching of special programs designed to decrease the opportunities for committing unlawful acts and to increase the risk for the criminals
- Support for the research efforts directed toward crime prevention
- Appropriate measures taken to promote cooperation

During the past decade, Hungarian representatives of the criminal sciences responded with sensitivity to popular demands for crime prevention. They explored the social causes of criminality. The importance of studying these issues is reflected in the many conferences, research programs, and published monographs.

HISTORICAL OVERVIEW

The police, with the consent and support of the Ministry of Interior, the Ministry of Justice, and the organizations of state administration, attempted in 1974 to impress upon the government the necessity for crime prevention. The main reason for this was the fact that a 1974 Council of Ministers Decree declared crime prevention to be a basic task of the police. Crime prevention in those days manifested itself mainly through coercive measures taken by the police authority, often in violation of basic human rights. The police attempted to take on the social education of youth, the settling of conflicts among ethnic groups, and the managing of social problems related to alcoholism, but these measures led to no great success.

Today, the regrettable fact is that an even greater increase in crime and other deviance is projected. Part of this assessment stems from the fact that there will always be perpetrators who consider crime a normal source of livelihood and will pursue a criminal lifestyle. Their conduct presents a permanent threat to the community and its law-abiding citizens. The police feel responsible for the prevention of crime. They believe it is important to systematize their knowledge and information, and they feel responsible for developing and implementing the preventive measures falling within their competence.

The establishment of special crime prevention units within the police force took place between 1989 and 1991. During this time, the Crime Prevention Department of the Hungarian National Police Headquarters was established and an internal structure was set up. At the regional level, organizational changes emerged from the single, independent crime prevention desk officer who works alone to crime prevention at the department level. Unfortunately, for many long years there has not been a uniform structure.

THE CURRENT ORGANIZATIONAL STRUCTURE

In 1991, under Provision 16/1991, the High Commissioner regulated all the crime prevention tasks of the various branches of the police organization. This provision gave an overview of the crime prevention tasks of the police. Article 7 states that the tasks defined therein "shall be executed in a division of labor and in close co-operation with the relevant service branches." In practice, this was not carried out because various organizational, personnel, technical, and financial conditions that were indispensable were not met. Also, the regional and local police managers understood the provision as a collection of tasks for the crime prevention units only. The major shortcomings of the provision are that it undertakes tasks unrelated to activities of the police and that it does not specifically separate the tasks relevant to the individual service branches. This quagmire has allowed everyone to "push off" the tasks to other service branches.

The Three-Year Development Plan of the National Police of the Republic of Hungary, 1995–1997 dedicated a mere two pages to crime prevention. In these pages, it delineated the directions of development for the period. In spite of its brevity, the document precisely evaluated the situation and drew some rational conclusions. The development program scheduled a gradual, annually implemented structural reorganization.

The Crime Prevention Department of the Hungarian National Police Headquarters was established in 1995. It included four sections—two prevention-organization sections, one Child and Youth Protection Section, and one Evaluation and Information Section. The establishment of the structure was delayed and, as of March 1, 1997, the department was again reorganized and the staff was drastically cut back.

The establishment of the Crime Prevention Departments of the regional (county) police headquarters was postponed to 1996. Following the reorganization, a national survey was prepared on the personnel strength of the crime prevention units. This document demonstrates the fact that the otherwise-progressive High Commissioner's concept was not fully implemented. The contents of the provision were, on several occasions, not executed by the local organizations. In one county with an outstandingly high crime rate and a high percentage of socially disadvantaged citizens, the strength of the Crime Prevention Department at the County Police Headquarters was a total of five staff members—the Head of the Department, two desk officers, a civilian employee, and an international relations desk officer not responsible for substantial work within the department.

CRIME PREVENTION WITHIN THE CRIMINAL INVESTIGATION SERVICE

The establishment of a Criminal Investigation Division did not lead to a significant decrease in the number of crimes committed. Also, the increased number of patrol officers and various technical developments did not result in a decrease in the number of crimes or enhance the public's perception of security. What is more, they did not even elicit satisfaction with the police's performance.

The severity and extent of domestic crime in Hungary has not as yet reached the level of many western countries. That is why there has not as yet been a widespread outcry for action among the community. The country, however, should not delay launching new crime prevention tactics until pressure from the society forces the criminal justice system to take the initiative. It is important to learn from examples of the welfare societies and meet problems halfway, instead of waiting to experience, through enormous sacrifices, the same problems that other nations have faced. The organizations of criminal prosecution and criminal justice will soon (if they have not done so already) reach the upper limit of their efficiency, and cannot be further enhanced significantly by simply increasing strength and expanding the infrastructure.

Crime prevention and criminal prosecution in the future requires a closer joining of forces at the societal level. One must step beyond the opportunities and means of a penal code. One should seek other concepts and methods. Hungarians must create conditions, based on the analysis of cause and effect, that make abstention from crime desirable and possible. Unfortunately, it can also be said of Hungary that the relationship between the police and the citizenry has deteriorated.

Consequently, the sources of information that would be invaluable to successful criminal prosecution have narrowed. The relations between police and the public and between the police and various civilian organizations need to be changed in order to promote cooperation.

Uniformed police personnel patrolling the community have a significant role in maintaining public relations. Officers, by the very nature of their work, have day-to-day contact with the citizens. District police officers have primary, direct information on the customs and habits of individuals and groups of individuals living in their area of operation, and they can actively contribute to the shaping of the public's image of the police. The crime prevention service established after the change of political systems was perhaps the first police unit that, on a large scale, opened doors toward the civilian population. It was an inevitable step because of the nature and complexity of their activities.

Because crime prevention involves all people and every facet of society, however, the police should not become its exclusive custodians. The police can only undertake tasks that are commensurate with their legal province, professional profile, and capabilities. Not ignoring the mobility of criminals, we must be aware that perpetrators still tend to commit crimes in their own immediate environment. Consequently, crime is predominantly a local problem, that requires local prevention and the development and implementation of a local strategy. The immediate communities have fundamental opportunities in the area of crime prevention. Because the problems and tasks predominantly occur locally, they also can be resolved at the local level. It is at this level that the root of problems can be grasped and problems attacked through the cooperation of various institutions with support from the community. The most rational solution for the alleviation of crime may be the adoption of "communal crime prevention." Under such a program *a community* would

- Assess the crime and public security status
- Delineate the causes and circumstances of criminality
- Prepare a program or strategy for the elimination thereof
- Implement the program through involvement of the partners
- Check the efficiency of the program and define supplementary tasks

The main entities implementing this strategy primarily would be the government and social institutions and organizations, civilian organizations, and voluntary members and groups of the population who are able to participate in implementing well-defined programs, have appropriate competence, and are supported by the police.

Government Decree 1136/1997, on the preparation and implementation of the government's comprehensive crime prevention program, binds the police to the creation of an organizational structure that will successfully compete in the area of crime prevention. In the initial period, the police are to launch, encourage, and foster self-defense strategies for purposes of crime prevention; urge the parties involved to decide on what local action to take; and achieve the goals set by themselves.

An indispensable part of a new police crime prevention philosophy would be a shift in attitude. Officers should be sensitive and helpful when dealing with citizens' problems and rendering services to the community. The police should openly reach out to those government and private sectors which are able and willing to do something in the area of crime prevention. They must take the first steps if they want to win the trust and confidence of the citizenry. The elements outlined above contain the basic principles that determine the general course, tasks, and shape of crime prevention. These principles must be taken into account when developing local strategies and the concepts of the local organizations.

The Building Blocks of Police-Level Crime Prevention

Effective police-level crime prevention can become a reality only if several conditions are met. First, action must be based on a uniform concept coordinated with and accepted by all service branches. Also, its organizational structure must provide professional guidance while at the same time not interfering with the decisions of staff members. Third, an effective crime prevention unit must have an efficient, trained staff. Fourth, the execution of the identified tasks is a uniform system of regulations that designate the crime prevention tasks for all service branches in a compulsory manner, and sets forth the basic principles of shaping the professional priorities, the organizational structure of execution, and the forms of cooperation. Police-level crime prevention also requires support and assistance from other governmental and social organizations. The police must establish a fair system of cooperation with organizations, and provide them with wide professional assistance in the crime preventive effort. Fifth, at all levels of police training, the need for crime prevention, its tasks, methods, and achievements are incorporated in the curriculum. Finally, crime prevention requires a steady financial backing which is not tied to the occasional financial donations of civilian organizations and which would permanently ensure coverage of expenses arising from the execution of prevention tasks.

Police Reorganization for Crime Prevention

The crime prevention units of the Criminal Investigation Service work with set minimum-strength levels of active-duty police officers:

- Hungarian National Police Headquarters, in Budapest, Crime Prevention Supervisory Department (CPSD)—ten officers
- County-level (metropolitan) police headquarters (CPSD)—six officers

Subunits, groups, or desk officers at police stations are hired depending on the population of the jurisdiction as follows:

- Above 100,000 population—mandatory subunit of six officers
- Between 50,000 and 100,000 population—mandatory group of four officers
- Between 25,000 and 50,000 population—mandatory single, desk officer
- Below 25,000 population—staffed according to the discretion of the head of the police station

Because of the large population and complex crime situation of any capital city, Budapest requires the implementation of a unique police structure. As cited above, the Budapest Police Headquarters has established a Crime Prevention Supervisory Department with a strength of at least 10 active-duty officers. At each of the district police stations at least two active-duty police officers are employed as independent crime prevention desk officers.

The organizational restructuring presumably was intended to improve the level of crime prevention in the capital city, and enhance the efficiency and success rate of activities. At a local level, the effectiveness of operations would also improve. At the same time, during implementation of the services offered to the capital city, the opportunity to concentrate the best possible personnel would continue to exist.

The new organizational structure was implemented in steps. Under Phase I (completed December 31, 1998), the HNP headquarters Crime Prevention Supervisory Department was established and reorganized. Also, the crime prevention supervisory departments of county-level headquarters, the Crime Prevention Supervisory Department of the Budapest Police Headquarters, and at least one independent desk officer of the District Police Stations was scheduled to be in place. In Phase II (completed December 31, 1999), the crime prevention subdepartments and groups of the town or city police stations were established. Furthermore, in as much as only one desk officer was appointed in Phase I, a second independent desk officer at the district police stations in Budapest was appointed. Phase III (completed December 31, 2000) saw the appointment of the independent crime prevention desk officers of the town or city police stations. The crime prevention units are organized to reflect the professional branches established and preferred within the Crime Prevention Supervisory Department of the HNP Headquarters.

Several branches of the crime prevention unit are given priorities in their work:

- Child and youth protection (DATARE [Drug, Alcohol, Tobacco and AIDS Resistance Education], search for missing persons, paedophilia, child abuse, juvenile delinguency)

- Preventive protection of property (theft, burglary, vehicle theft, security of banks and financial institutions, supervision and control of private investigations)
- Prevention of drug-related crime
- Information and advisory activities, targeting all layers of the population (police-community, police-media, and police-ethnic minorities relations), and the expansion of cooperation with them

In addition to the above-prioritized professional branches, crime prevention extends, as an overall task, to analysis of the phenomena related to how people become victims and the evaluation of the prevailing situation of crime prevention.

COMMUNITY POLICING IN HUNGARY INTRODUCTION AND IMPLEMENTATION

The idea of community policing can often be found in literature on public safety and crime, and in the sense of security felt by the community. Following the dramatic change in the political and socioeconomic system in Hungary, we must rethink the police functions, roles, and activities. It has become obvious that the Hungarian Police can operate in an efficient way only if they enjoy the confidence and support of the community. Therefore, it is necessary to clearly delineate the specific responsibilities of the police and to ensure the proper training of the support staff in accordance with this challenge.

The police of the Hungarian Republic wish to operate in harmony with the requirements of the constitutional state and with changing social conditions. The new concept of public security means that the level of safety in a given community is the product of the community living in that area. Though the police have the primary responsibility for this task, local governments, prosecutor's offices, courts, various civilian organizations, safety services, and many other agencies are also responsible for the safety of a given neighborhood. A main objective is to create a police force that serves the community in cooperation with the people.

Among the professional responsibilities of the public safety service, a new model of practice is gaining emphasis. Instead of a punishment-centered view, the main objective is the implementation of service-oriented patrolling built on a systematic communication with people in the neighborhood. In the course of getting this under way, the service-oriented patrolling scheme is faced with four main challenges:

1. Community-oriented prevention of crime and accidents at the national, county, and local levels

2. Reorientation of the patrol service, focusing on sympathetic behavior toward victims and a service-oriented attitude instead of punishment-oriented behavior
3. Decentralizing the management of the entire organizational hierarchy, since this is the only way for the staff to ensure greater freedom for their activities
4. Instituting civil control of the police, police accountability, and the involvement of citizens in setting priorities

Improving the public safety units of the Hungarian Police involves adopting this new model of patrolling. In January 1997, the Hungarian Police Chief stated that "the objective of the patrol service is to create [a clear] police presence in public places, and to build up co-operation with the law-abiding community." Professional regulations also require that all officers show a high level of moral behavior, in order to gain the confidence and co-operation of law-abiding citizens.

Regarding cooperation, it should always be kept in mind that the parties involved have separate duties, responsibilities, and fields of authority. These responsibilities cannot be delegated. This means that the legal background should constantly be monitored and modified in accordance with changing conditions and requirements.

Several efficient programs and organizational structures can be successfully utilized in the implementation of community policing. In 1953, the so-called local police service was established, meaning that a uniformed officer started working in a given locality. Regulations concerning the activities of these officers made it possible for them to organize and coordinate the security of the local community. At present, there are more than 2,000 such officers in Hungary. Their activities can be improved to enhance community policing.

According to various foreign experts, the main weaknesses of the Hungarian Police are the lack of citizen confidence, a very low level of good relations with the community (in particular with ethnic minorities), and ignorance on the part of the police regarding the importance of their own accountability.

Foremost in the mind of the average citizen is concern over public safety. In harmony with community requirements, police activity should focus on public security that is built on cooperation. Community-oriented policing is based on the fact that the solution to social problems and crime management should be carried out in close connection with the civil sphere. This is the only way to guarantee a fair and efficient police service. One of the most important elements of community-oriented policing is that both the police and citizens understand the local conditions and the problems to be solved. Consequently, they are able to formulate and implement programs together. Evaluation of police services should be done by the community who require these services.

International cooperation offers an excellent chance for learning from the experience and methods of other nations that can be utilized in Hungary. One of the essential elements in implementing the objectives is to make the Hungarian Police staff learn about the philosophy of community policing. To that end, training courses between May 1998 and May 1999, with the help of the Police Research Institute of Louisville University (Kentucky) and the Project for Ethnic Relations Foundation, were implemented. The main objective of this program was to introduce the philosophy of community policing. It was also intended to change the motivation of the officers and facilitate the democratization of the Hungarian Police, with particular emphasis on human rights and freedoms.

Based on the experiences from the courses, the next step was to organize systematic professional training for the staff. The publication of a new professional booklet proved very useful. Lecturers participating in the training course were able to use their new experiences in their work at the police training schools.

SUMMARY

As noted earlier, Hungary can and must face the regrettable fact that a further increase in crime and other deviance is projected. Crime prevention in the future requires a closer cooperation with the community. Because crime is predominantly a local problem, it requires local prevention and the development and implementation of a local strategy. Community-oriented policing may prove to be the best approach for enhancing the efficiency of the police in fulfilling this objective.

CHAPTER 13

CRIME PREVENTION: A COMMUNITY POLICING APPROACH IN RUSSIA

Yakov Gilinskiy

In a "war on crime" the use of repression alone is no panacea. Putting one's faith in a policy of crime prevention would be far wiser. It is also clear that crime prevention, especially community crime prevention, is impossible without the active support of the police, particularly in the form of community policing. The goal of the police in a war on crime must refocus on the defense of society through community crime prevention.

The main obstacles to achieving this goal include the legal nature of community crime prevention and community policing, the mechanisms of interaction between the police and the public, limitations on interventions into private life, and the lack of efficiency of crime prevention. To address these tasks and problems in Russia, this chapter focuses on the present situation in Russia; briefly addresses basic theory on crime prevention; outlines the concept, history, and reality of crime prevention in Russia; and discusses contemporary mechanisms and the future of crime prevention.

CONTEMPORARY RUSSIA IN BRIEF

Russia, or the Russian Federation (RF), came into existence in 1991 after the breakdown of the Union of Soviet Socialist Republics (USSR). Under the USSR, what is now Russia had been labeled the Russian Soviet Federated Socialist Republic. The Russian Federation is over 17-million square kilometers in size (greater than 6.5-million square miles). It consists of 21 republics, six Krai (individual administrative territories), 50 provinces, one autonomous area (Chukotsk), and two cities under federal administration—Moscow and St. Petersburg. The population of the Russian Federation has grown from 102.9 million in 1951 to 148.2 million in 1996. Seventy-eight percent of the population live in the European part of the country and 22 percent in the Asiatic sector (west Siberia, east Siberia and the Russian Far East). The population is 47 percent men and 53 percent women. In 1995, 73 percent of the people were living in urban areas and 27 percent in rural areas (Russian Statistic Annual, 1995:17).

The population is composed of a variety of ethnic groups. Russians make up 83.5 percent of the population, with Tartars composing 3.8 percent, Ukrainians 2.5 percent, and Chuvashians 1.1 percent. Each of the other groups represents less than 1 percent of the total population.

Social, Economic, and Political Situation

Since the breakup of the USSR, Russia has faced serious challenges. Gorbachev's *perestroika* was a necessary attempt to save the power structures through a process of reform. Khrushchev and others have made similar efforts in the past. Each, however, has ended with the actual or political death of its propagator, and has been followed by a period of stagnation.

Gorbachev's reforms turned out to be the most radical, although many of these (such as *glastnost,* the multiparty system, the release of states occupied by Stalin, the lifting of the Iron Curtain, and the granting of the right to hold private property) did not turn out to be fully satisfactory. Even today, symptoms of socioeconomic catastrophe remain untreated. Power is retained by the ruling class. Corruption, commonplace in Russia, has grown monumentally in all establishments and organs of power. The militarization of both economics and politics continues. Interethnic conflicts have given rise to mass murders. Nationalist, anti-Semitic, and neofascist groups have formed and met with no resistance. The war in Chechnya is terrifying evidence of the neototalitarianism.

The ever-growing economic polarization of the population, visible in the stark contrast between the poverty-stricken majority and the "new Russians" (a criminalized, nouveau riche minority) is a source of very real social conflict. The difference between the incomes of the 10 percent least prosperous and the 10 percent most prosperous was 1:4.5 in 1991, 1:8 in 1992, 1:10 in 1993, and 1:15 in 1994 (Financial News, 1992; Social and Economic Situation in Russian Federation, 1994:139). However, there are hopeful signs. For example, the economic reform now under way in Russia (that is, the transition from a planned state-run economy to a market economy) is, beyond a doubt, progressive in nature.

The redistribution of property is occurring both legally and illegally (the latter accompanied sometimes by bribery, murder, or threats). Technological backwardness and the incompatibility of the native production and service spheres have manifested themselves in the course of reforms. As a consequence, the staff of industrial plants and enterprises appear inferior—unqualified and marginalized. The disintegration of the services sphere and of the social infrastructure have caused further difficulties for the population. Moreover, a great number of people are not paid for long periods of time (often many months). The endeavors of science, education, medicine, and the arts have no support, and struggle to exist, much less progress. Finally, the government permits mass human rights abuses, particularly in military and penal institutions where tyranny and torture dominate. Confirmation of this travesty of justice can be found in international research compiled by Amnesty International and in documents generated by Russian writers (Abramkin, 1996). It is not surprising that the socioeconomic and political situation in the country has led to a growth in crime and other types of deviant behavior.

Crime in Russia

The incidence of crime in Russia has soared from 987 per 100,000 population in 1984 to 1,618 in 1997 (Crime and Delinquency, 1992; Crime and

TABLE 13-1 TRENDS OF CRIME IN RUSSIA 1985–1997 (INCIDENCES PER 100,000)

	1985	1987	1989	1991	1993	1995	1997
All registered crime	989	817	1098	1463	1888	1863	1551
Premeditated murder (with attempts)	8.5	6.3	9.2	10.9	19.6	21.4	18.7
Grievous bodily harm	19.9	13.9	25.0	27.8	45.1	41.7	29.8
Theft	324.7	251.1	512.1	837.3	1065.2	924.6	681.9
Robbery	29.9	21.0	51.0	68.8	124.3	94.7	69.9
Assault	5.8	3.9	9.9	12.4	27.0	25.4	21.3

Source: Crime and Delinquency (1992 and 1997).

Delinquency, 1997; State of Crime in Russia—1997, 1998). A brief decline between 1986 and 1987 was due to the influence of positive social and political changes that *perestroika* had had on the consciousness of the people. Following this honeymoon, various social and political crises caused the crime rate to rise again. The increase in violent crime is particularly significant. The rate for premeditated murder (with attempts) rose from 8.5 in 1985 to 19.8 per 100,000 in 1997. Similarly, the rate for serious bodily harm increased from 19.9 in 1985 to 329.8 per 100,000 in 1997. Such an explosive growth in the level of violent crime reflects the severity of the social crisis. Data from medical statistics are more dramatic. The rate of death by murder in 1990 was 14.3. In 1994 it was 32.6 (Russian Statistic Annual, 1995). The growth rate of robbery also accelerated—21.0 in 1987, 127.3 in 1993, and 75.6 in 1997. Assault increased by 6.5 times between 1987 and 1997 (3.9 to 23.1) (see, Table 13-1). The fall in the crime rate during the period 1994–1997 was not due to real positive change, but to "new" policies adopted by the police authorities that work toward a mass coverup of the registration and recording of crimes. In St. Petersburg, according to the results of a survey of victims, 12 percent of the respondents reported having been victims of crime in 1991, but by 1994 the figure stood at 26 percent and in 1995 it was over 30 percent.

THE RUSSIAN POLICE AND THE JUSTICE SYSTEM

The organization of the police of Russia, known as the Militia, was first set up a month after the state coup of October 1917. The current directives, functions, and structure of the Militia were laid down in Russian legislation entitled "On the Militia," passed on April 18, 1991. The Militia operate under the auspices of the Ministry of Internal Affairs (MVD) of the Russian Federation, along with the internal army, specialized police forces (for example, railroad, air, and river police), the fire safety services, penitentiary personnel, and others. The Decree of the President of the Russian Federation transferred the Russian penitentiary system from the MVD to the Ministry of Justice. The internal army is responsible for dealing with internal conflicts, rebellion, disturbances, and riots. The directives of the Militia are to provide citizens with personal safety, to stop and prevent crime and civil law breaking, to solve crimes, and to secure civil order and safety within society. The size of the Militia in 1995 stood at about 540,000, and that of the internal army at about 278,000.

The Militia is organized into two main subdivisions—the Criminal Militia and the Militia for Civil Safety (that is, public order) at the local level. The Criminal Militia includes the Detective Service, the Economic Crime Prevention Service, scientific-technical specialists, operational investigators and others who supply material for criminal investigation, and the Economic Crime Prevention Services. The civil Militia include the Duty Service, the Service for Securing Civil Order, the State (government) Automobile Inspectorate (GAI), the Security Service, divisional inspectors, temporary detention guards, the crime prevention service (which includes the Inspectorate for dealing with juveniles), and various other departments. The Criminal Investigation Service is a separate unit under the Ministry of Internal Affairs.

The Militia is given far-reaching powers, including the right to enter—without a warrant—the living places and other premises of citizens, the premises of companies, organizations and official departments (excluding those of foreign diplomatic representatives), and to conduct searches on transport facilities and to search not only the baggage but the actual person of citizens. The law "On the Militia" regulates the terms and procedures for use of physical force, special methods (rubber truncheons, tear gas, water cannons, armored cars, and others), and firearms. The specific principles guiding the actions of the Militia are supposed to be lawfulness, humanism, respect for human rights, and *glasnost* (openness or transparency). As with restrictions of force, Militia actions often violate all of these principles, with concrete examples of such violations being regularly reported in the Russian and foreign media.

THEORY OF CRIME PREVENTION: A SHORT SURVEY

Crime prevention is one of the elements of social control over criminality. Social control is the mechanism of self-organizing and self-preservation of society by the establishment and maintenance of a normative order. Two basic regulators of individual behavior produced by society are social values and the norms appropriate to them. Both values and norms are transferred by various methods. Two basic methods are encouragement and punishment.

There are various forms of social control (for example, formal and informal, internal and external, direct and indirect) and numerous patterns (Black, 1976; Davis and Anderson, 1983). Different social institutions (from the family and the school to the police and prison) carry out the functions of social control. In general, social control can be reduced to the fact that society, through institutions, sets values and norms, provides for their transference by socializing individuals, encourages their observance, and punishes their infringement.

Social control over criminality often includes a call for a "war on crime," armed with both reprisals and crime prevention. Society has tried all means of reprisal, including the death penalty and torture. Criminal behavior, however, has not disappeared. At present, the conventional wisdom holds that there exists a "crisis of punishment," that is, a crisis of criminal justice and a crisis of society's control over criminality, including the control of police (Albanese, 1990; Barkan, 1997; Davis and Anderson, 1983; Donziger, 1996; Hendricks and Byers, 1996; Rothwax, 1996; Sumner, 1994). The National Criminal Justice Commission of the United States advocates shifting "crime policy from an agenda of 'war' to an agenda of 'peace'" (Donziger, 1996:218). Barkan (1997:542) advises society to "reduce reliance on imprisonment and to put more emphasis on community correction."

Modern western policy for the social control of crime includes

- Recognizing the irrational and inefficient use of reprisals ("crisis of punishment")
- Changing the strategy of social control from "war" to "peace" or "peacemaking" (Donzinger, 1996:218; Pepinski and Quinney, 1991)
- Searching for alternative (nonrepressive) measures
- Giving priority to crime prevention

Crime prevention is understood as the combined influence of society, institutions of social control, and individual citizens on the causes of crime that result in the reduction of criminal behavior, leading to desirable changes in the social structure and to the prevention of potential crimes. The priority of prevention was precisely stated by Montesquieu and then repeated and advanced by Beccaria (*On Crimes and Punishments*). Voltaire has named the prevention of crime "true jurisprudence."

In modern literature, three levels of prevention are distinguished. Primary prevention entails influences on the environment, ecology, economics, and sociopolitical conditions of life with the goal of improving those conditions (the Russian term is "general prevention"). Secondary prevention involves the maintenance of security measures, influences on "groups at risk," and the elimination of circumstances that encourage crime (the

Russian analogue is "special prevention"). Tertiary prevention is "individual prevention" (Russian criminology). The whole concept of crime prevention is much more reasonable, more democratic, more liberal, more progressive, and certainly more pleasant than "struggle" and reprisals.

But, is crime prevention feasible and efficient? Attempts to prove efficacy raise several other issues and problems. First, what is the object of prevention, if many criminologists do not know what such "criminality" is? Second, prevention influences the causes of crime and circumstances that engender crime. But who really knows what these causes and circumstances are? Third, it is not surprising that there are no convincing data on the efficiency of different prevention activities. In one analysis, Graham and Bennett (1995) assemble a large amount of material about different prevention programs. However, they do not prove the effectiveness of the programs. Finally, there is serious danger of the degeneration of prevention into the infringement on elementary human rights. Steinert (1995) compared the "instrumental rationality" of prevention with Auschwitz, and wrote in 1991, "I see the whole idea of prevention as part of one of the grave mistakes of this century" (see, Albrecht and Ludwig-Mayerhofer, 1995:5).

Despite the foregoing criticism, one cannot totally discount the feasibility of prevention. In fact, there are several reasons for its support. First, the processes of organizing and stabilizing the daily life of a community are prime objectives for society. Second, society will, by and large, react to crime, and prevention is always preferable to after-the-fact reprisals. Third, the set of primary, secondary, and tertiary prevention measures should improve social conditions, create a more humane atmosphere and, as a result, serve to reduce inhuman actions. Finally, secondary and tertiary prevention measures are capable of protecting, especially at the community level of crime prevention, specific persons and potential victims, and of rescuing them from possible encroachments.

CONCEPT AND HISTORY OF CRIME PREVENTION IN RUSSIA

Marx and Lenin have repeated the words of Montesquieu about the priority of crime prevention in comparison to punishment. The priority of prevention is part of the ideology of Marxism-Leninism. But how the concept was realized in practice is another story. Khrushchev reanimated the idea of prevention, and at the Twentieth Congress of the Communist Party of the Soviet Union (CPSU) in 1956, he advocated making crime prevention a priority. This directive was repeated at the Twenty-first Congress in 1959: "It is necessary to undertake such measures, which would prevent and then completely exclude the occurrence of any offenses, any harm to society. The main idea is prevention, upbringing." The program of the CPSU was accepted at the Twenty-second Congress in 1961, which declared that major attention "should be paid to prevention of crimes." Khrushchev saw in crime prevention the panacea for "antisocial" activity. He believed in the efficacy of prevention and promised to shake the hand of the last criminal in the USSR.

A liberalization of punishment was observed during the "thaw" when Khrushchev reduced the number of persons condemned to imprisonment (the majority were transferred to labor collectives for "reeducation." The rate of registered criminals (per 100,000 population) was reduced 19 percent between 1961 and 1965 (Crime and Delinquency in USSR, 1990:12). The priority of prevention was cited in all the party documents. At the Twenty-fifth Congress of the CPSU in 1976, the idea of a complex approach to the prevention of offenses was proclaimed. But, with the ending of the thaw, repression of criminal and administrative justice returned, and the overall crime rate, suicides, alcoholism, and drug use began to increase.

New positive tendencies were observed during the period of Gorbachev's *perestroika.* The number of persons sentenced to imprisonment decreased, and the number released from criminal liability with the application of alternative measures of "public influence" increased (Crime and Delinquency in USSR, 1990:12). At the same time, the crime rate fell by 16 percent between 1985 and 1987. The modern period (post-*perestroika*) is characterized by a return to a repressive policy and the growth of all kinds of criminal behavior, suicides, alcoholism, and narcotics use.

The party declarations were demagogic and the idea of complete "extermination" of criminality in "socialist" ("communist") society was utopian. But those ideological clichés allowed scientists to develop theoretical and methodological bases of prevention, and to put them into practice.

The view of western criminologists that, in the USSR, there existed only two concepts of the causes of criminality—"the relics of capitalism" (relic theory) and the "influence capitalist of encirclement" (influence theory)—is not valid. While those were the basic ideological and propaganda positions of the CPSU and the Soviet state, other modern sociological concepts were developed. Domestic criminologists tested the influence of Hegelian-Marxian theory, positivism, and interactionism, while developing the ideas of the Russian sociological school in criminology. The influence of Hegel and Marx, prerevolutionary criminology, and the hard social conditions in the Soviet state lead to the convergence of domestic criminological works with western critical, radical, and "structural" theory.

One stimulus for the development of crime prevention in the former USSR was recognition of the complex socioeconomic planning and development of labor collectives. The idea, born in the 1960s in Leningrad, was "approved" by leadership of the Leningrad Regional Committee CPSU and was widespread across the country. For CPSU leadership, it was an attempt to maintain the system by adding to the centralized planning of economic production—the planning "from the roots up" of social development (for example, improvements in health care, working conditions and modes of life, adequate housing, "communist education of workers," and "prevention of criminality").

The complex socioeconomic planning provided scientists with an ideological "roof," allowing them to carry out empirical study. For example, empirical criminological research was carried out in Leningrad, Orel, Murmansk, and other regions with the purpose of elaborating "the scientifically reasonable plans of complex socioeconomic development," including crime prevention (see, Spiridonov and Gilinskiy, 1977). Under the same ideological roof and technique of socioeconomic planning, planning for crime prevention in labor collectives and "on a residence" (that is, community crime prevention) were developed. As a result, in the 1960s and 1970s, the theoretical and methodical bases of crime prevention were developed.

The basic approach to prevention was to identify the causes and conditions of criminal behavior, for the purpose of neutralizing or eliminating them. This stimulated research on the causes and factors influencing the conditions, rates, structures, and dynamics of crime and various kinds of crimes (Kudrjavcev, 1968; Karpetch, 1969; Jakovlev, 1971; Spiridonov and Gilinskiy, 1977). The three levels of prevention were general social (primary prevention), special, criminological (secondary prevention), and individual (tertiary prevention). There were also geographical levels of prevention focusing on the republic, region, area, city, labor collective, and community ("a residence"). The geographical aspect of criminal behavior has been studied by Gabiani in Georgia, Leps and Raska in Estonia, and Avrutin and Gilinskiy in Leningrad.

The topics of prevention were classified according to the description of their purposes, problems, functions, and methods of activity (for example, party committees, soviets of people's deputies and their executive bodies, law enforcement bodies, labor collectives, public organizations, and citizens). There were also organizational bases of prevention, including information provision (empirical sociological and criminological study), forecasting of criminal offenses, planning and coordination of preventive activity, and analyses of results and estimation of efficiency.

A prevention service was created in the Ministry of Internal Affairs (MIA). Additionally, an independent course in prevention was introduced in educational institutions of the MIA. Plans for crime prevention were developed at all "geographical" levels. Voluntary public patrols (VPP), *komsomol* operative groups, public (communal) courts, and councils of prevention in labor collectives were all functioning. District bases for public order guards (DBPOG) were organized and worked in the communities.

The bases included police inspectors, officers from the inspection of juvenile delinquents, members of VPP, and members of public courts. Thus, there was an attempt at crime prevention.

The prevention activity was not without its shortcomings, however. First there was a gradual degeneration and regeneration of initial VPP in a system of formal "measures" for reporting and in specialized formations known as *comsomol* operative groups. Second, some prevention activity reflected an excess of power that infringed on human rights. This was true especially of the members of VPP and the public courts. Third, intrusions into the personal life of citizens often lead to their public disgrace due to the use of "the panel of shame" or "the wall newspapers" (that is, the public posting of names). These and other problems can be attributed to the squalid conditions endured by people residing in urban "communal flats" with their never-ending quarrels, conflicts, and the mass bitterness of the population.

EXISTING CRIME PREVENTION IN RUSSIA

The realization of crime prevention in Russia was pushed to the background at the beginning of *perestroika* and was superseded by political, economic, financial, and ideological problems. The prevention service was liquidated by the Minister for Internal Affairs (MIA), and only later was it restored. The study of prevention in MIA educational institutions was canceled. Eventually, it became a special part of criminology. The majority of the former types of crime prevention (voluntary public patrols, public courts, the councils of prevention in labor collectives) ceased to exist.

The decrease in the crime rate during the years of *perestroika* (1985–1987) was replaced by major increases since 1989. The increase lead, partly as a result of fear of crime, to "moral panic" (Cohen, 1996). The reaction on the part of law enforcement bodies, headed by the MIA, was to "strengthen the struggle." As a result, the number of people imprisoned in Russia increased to 760 per 100,000 in 1996 (first place in the world, with the United States occupying second place at a rate of more than 560 per 100,000). Certainly, this did not contribute to the eradication of crime.

Attempts to revive the old forms of prevention under new conditions are doubtful. Thus, the VPP today can become a legal cover for radical groups, including neofascists (this assumption has become a reality: see, *Izvestia* [News], September 23, 1997). Communal (public) courts have not reappeared. It is more expedient to form new "arbitration public courts." Employees act as intermediaries between the offender and victim and carry out restorative justice. In terms of community policing, domestic regional police of the public order (Militia for Public Order) are trained and follow the methods of the federal criminal police.

On the other hand, there are some effective forms of community crime prevention. Public organizations concentrating on "self-help" have emerged for drug addicts, prisoners and former prisoners, homosexuals, and others. Public and commercial organizations offering social and psychological help assist victims of violence, teenagers, female victims of family and sexual violence, and suicide. Neighborhoods have formed associations such as Neighborhood Watch programs and private security groups. Although these steps are no panacea, they represent visible signs of success.

Russia now has a paradoxical situation. The normative documents dictate the prevention of crime and delinquency (for example, the Order of the Minister for Internal Affairs, August 6, 1993, "On Work of Services and Subdivisions of MIA on Crime Prevention"). However, no action is taken on these directives. At the same time, diverse regional initiatives do exist, including voluntary public patrols, district bases for public order guards (DB POG), Cossack's societies, public organizations (such as Social Health and Legal Order), and others (see, Kosoplechev and Izmailova, 1997).

MECHANISMS OF EXECUTION OF CRIME PREVENTION PROJECTS

The successful execution of diverse projects, programs, and plans runs into continual difficulty in Russia. It is a "tradition" inherited from the former Soviet Union that a great number of programs and plans are outlined but nothing happens. For instance, the Federal Program of the Russian Federation on the Combat of Crime (1996–1997) and many regional crime prevention programs (including those operating in St. Petersburg) were elaborated but not acted upon. Certainly, elaborate dreams and plans are easier to evoke than to bring about. Real mechanisms of program execution, especially plans and funds for their realization, are absent. Scientists from the Research Institute of the Prosecutor's Office of the Russian Federation wrote,

> Systems of prevention . . . were destroyed. . . . In fact no structure of power controls the situation of crime prevention. . . . The programs of combatting crime are completely unrealized. . . . Financial means are absent (Kosoplechev and Izmailova, 1997:84,86).

Moreover, the Russian citizens have given up hope for crime prevention by the power of the police. Civil initiatives are viewed as more efficient. For example, groups of citizens often organize their own security programs in Moscow and other cities. Such programs are analogous to Neighborhood Watch.

THE FUTURE OF CRIME PREVENTION IN RUSSIA

The prognosis for the success of crime prevention in contemporary Russia is very difficult to predict because of the unsteady economic, social, and political situation. On the one hand, continual economic, social, political, and moral problems are more important than problems of crime prevention. If Russia has no money for health services, education, science, culture, or social welfare, then there is certainly no money for crime prevention. Moreover, the Russian police prefer using power as the primary means of combatting crime.

On the other hand, crime is a serious problem. Coercive power alone is not efficient. Theoretical and methodological bases and experiences with crime prevention are present in the country. Fear of crime stimulates public activity and public organizations for "self-help" and Neighborhood Watch programs. Conditions generated by a dubious criminal policy, crises of authority and criminal justice, the corruption of state and law enforcement bodies, and unskilled efforts from the general populace have prompted a general outcry for crime prevention. International contacts with scientists and practitioners from other countries should prove very productive.

SECTION 4

This final section of the book features chapters on Mexico and China. Both nations maintain strong centralized authority over the police and the agencies of social control. However, each country differs greatly in its social milieu. Mexico's policing and crime prevention initiatives can be found in various national plans and programs. Much of the activity revolves around attacks on organized crime and drug trafficking. The direction for these endeavors comes from the national government. Unfortunately, many initiatives face stiff opposition in the countryside where organized drug and crime groups are well established and tolerated. Concurrent with these policing initiatives, there is a strong emphasis on helping families, educational institutions, and communities to promote morals, values, and ethics among the populace. It is believed that strong social control must come at the local level through efforts other than those of official, formal policing agents.

Whereas Mexico is now seeking stronger, informal community control, China has a history that has always embraced this ideal. Community policing and crime prevention in China can be viewed as a simple extension of age-old social practices. The family and local community have maintained primacy in dealing with deviant behavior. The central government has encouraged this activity, instead of imposing formal social control agencies. One of the current challenges involves how to maintain this traditional system in light of the steady growth of cities and the influx of outside influences. More formal methods of control, such as the police, are becoming increasingly necessary. Thus, China is trying to maintain a system of informal social control that Mexico is only now attempting to develop.

CHAPTER 14

CRIME PREVENTION POLICIES AND CIVIC MORALS IN MEXICO

Walter Beller Taboado

Social studies based on a theory of open systems sharply contrasts with both a social individualist, or atomist, approach (by which persons can, in principle, be studied in a vacuum), and a holistic, or global, interpretation, which analyzes all phenomena by having recourse to an abstract notion of totality (Bunge, 1983). Curiously enough, individualism is not appropriate for the study of many social problems, such as drug crimes, because it considers individual malfunctions apart from the cultural, economic, and ideological components that lie at the root of people's psychosocial identities. Holism alone is also not adequate, for it takes into account socially anomalous individual behavior without recognizing that it is free and autonomous individuals who assimilate and process social inputs.

By contrast, the systemic approach views society as a group of individuals related to one another by sets of activities and institutions. In the systemic vision, some properties are indeed strictly individual, but others are emergent; that is, they spring out from the whole of an individual's links to society. In general, a property is emergent if isolated individuals do not possess it. This is clearly the case, for example, of the property of "being a victim." In order for a man to become a victim, there must be another person who makes him a victim. People neither are nor are not victims. There is no victim without victimization, and the other way around. Thus, "victim" indicates being an emergent property.

The systemic study of the drug-crime relationship includes several levels of organization and integration. The very idea of levels of organization and integration derives from biology and has to do with the fact that the organization of life may be discerned at different levels of complexity, from cells to ecosystems. For systemic analysis, therefore, there would be a first level of organization in which neurophysiological and psychological factors related to the addictive and criminal behavior would be analyzed. Another level of analysis would correspond to the legal consequences of such behaviors. A third, wider organizational level would concern the different sorts of effects that illicit drug trafficking has on the national economy, politics, security, and culture. One single phenomenon may be viewed from different (and complementary) perspectives.

A theoretical danger arises in thinking that the various levels of organization and analysis can be considered on their own (that is, in isolation). Such an approach leads to reductionist sorts of explanations, which are obviously unsatisfactory, and therefore to "isolated programs," which are bound to fail. On the other hand, the systemic approach considers the issues in which we are interested by using two strategies, namely, approaching crime from the bottom up and from the top down. This view presupposes a multi- or interdisciplinary kind of research, which faces two rather big problems. The first one is that each scientific discipline is studied on its own, independent of any other. In the academic world, the different sciences have their own compartments. Thus, although serious research tends to be more and more specialized, it is also true that multidisciplinary studies are in greater and greater need. It is important to learn how to put together different kinds of results and to involve sciences from different fields with different levels of specialization.

The second shortcoming has to do with the difficulties of interdisciplinary research. It is not enough to establish a common language. It is necessary to link concepts that come from a diversity of fields, such as the biological sciences, chemistry, social psychology, and law. This is a challenge for people working on global problems, and upon whom solutions of great importance for the stability of countries depend (Bunge, 1995). It is obvious that national and international programs have important socioeconomic, political, and moral foundations whose lack of coherence can be counterproductive. Underdeveloped countries face problems of poverty (often extreme), which are practically unknown in developed countries.

Thus, it can be argued that the systemic approach is a kind of synthesis of holism and individualism, since it preserves the notion of the totality of global properties while holding to the study of the individual's features. What matters is to establish how systemism could help in the diagnosis and design of alternative policies for crime prevention.

NATIONAL POLICY TO FIGHT CRIME

Mexico's federal government has faced the crime phenomenon with a global strategy based on strict respect for the law and a desire to increase efficiency in its war against crime. The task is carried out by means of an integrated, interagency approach, which constitutes the foundation of crime prevention and drug dependency policies and makes them functional.

Article 26 of Mexico's Political Constitution states that it is one of the government's functions to organize and implement a national development plan. This broad program seeks to instill stability, dynamism, permanence, and equity in the country's economy. Its ultimate goal is to make the nation politically, socially, and culturally democratic and independent.

In accordance with an explicit pronouncement of the legislative branch, the democratic planning system takes into consideration the desires, demands, and needs of society as a whole. These goals should be incorporated into the National Development Plan and into the more specific programs to which it gives rise. A central goal of the plan is to enact a set of legislative tools that would regulate relations between rulers and those ruled, paying special attention to the social and political plurality that exists in Mexico.

Complying with the aforementioned constitutional provision and with the stipulations of the Planning Law, the present federal government formulated the National Development Plan for 1995–2000. This plan is the outcome of a broad opinion survey of the entire country's population. In the area of law enforcement and the administration of justice, the most urgent issues were the strengthening of the rule of law, the war against impunity and organized crime (basically, but not only, drug trafficking), the eradication of corruption among public servants, professionalization (so as to establish a civil service) and improvements in the provision of public services. Thus, one of the main goals was to guarantee that law enforcement and the administration of justice will be quick, efficient, and respectful of civil rights.

At the federal level, the office of the Republic's Attorney General has, among other faculties, the power (and the obligation) to prosecute federal crimes and to intervene as a permanent part in *amparo* proceedings (*amparo* is a legal procedure, originally conceived in Mexico but now internationally known), which foster strict adherence to law and to the protection of public interests. The Attorney General represents the federation in all those matters in which he takes part. He intervenes, by means of nonbinding judicial opinions, in controversies in which diplomats and consuls are involved, in the promotion and execution of agreements of international scope in areas of police and judicial cooperation, in the international extradition of criminals, and in the application of treaties related to the international exchange of sentenced prisoners. The Attorney General also participates in constitutional controversies and challenges, as specified in Article 105 of the Mexican Constitution.

In accordance with the National Law Enforcement Program for the years 1995–2000, the office of the Attorney General of the Republic was assigned the following goals:

1. To reach optimum efficiency levels in criminal prosecutions
2. To create conditions that would allow the attainment of tangible results in the war against drug trafficking, by collaborating and cooperating at both a national and an international level
3. To generate professional and ethical conditions so that the federal judicial police and federation public prosecutors duly perform their duty, which is to guarantee the population's security
4. To ensure that the principle, according to which no one is above the law, is actually applied, regardless of the offender's economic, social or bureaucratic level

In order to achieve the goals set by the National Law Enforcement Program 1995–2000, a series of priorities were established for the office of the

Republic's Attorney General. Naturally, the organization of crime prevention activities stood out as crucial. The most important priorities stated were

1. The furtherance of agreements among federal, state and municipal governments (into which Mexico is divided) for the execution of joint actions to prevent criminal behavior
2. The promotion of social communication campaigns that would help in guiding the population with respect to preventive measures and in making public or explicit their rights in case they become victims of a crime (assault, rape, etc.)
3. The establishment of citizen participation programs in crime prevention, as well as of channels enabling the population to timely file complaints about illegal acts

The prevention of criminal behavior is as important as the actual war against crime. Prevention requires the cooperation of diverse government agencies and the work of civil organizations interested in preserving public security and social integration. As is widely acknowledged at both an international and national level, no preventive action can succeed without the constant and resolute participation of the community, in coordination with the authorities. On the other hand, criminal law comprises three dimensions: repression, prevention, and redress. Ideally, the first dimension should be applied only when the other two have failed. Since Beccaria's times, people have always been inclined to think that it is more preferable to prevent than to punish.

With respect to the issue of drug trafficking and drug dependency, the Mexican legal system seeks to attack the phenomenon in all its complexity. Criminal, toxicological, psychological, medical-legal, sociological, criminalistic, educational, administrative, and even fiscal (in the case of money laundering) specialists all work jointly in order to focus on the phenomenon as a whole. It is on this basis that the federal government has implemented the National Program for Drug Control 1995–2000. In this program, federal agencies in charge of educational and health services play a crucial role. Naturally, the office of the Republic's Attorney General, which deals with the issues of the prevention and control of narcotics, also takes part, along with other government agencies, such as the Ministry of the Interior, the Ministry of National Defense, the Ministry of Finances, and the Ministry of Communication.

To put it briefly, the fight against criminal behavior and its prevention are viewed in Mexico in an integrated and systematic way, in which the cooperation of both public and private organizations and the participation of civil society are viewed as essential.

PHILOSOPHICAL APPROACH TO CRIME PREVENTION

From a strictly legal point of view, crime prevention is nothing but a subsystem of the law enforcement system. It incorporates, however, two complementary aspects—the threat that the person who breaks the law will be investigated and punished, and the shaping of a moral awareness in the population that helps deter, prevent, and marginalize situations that oppose legal order.

A global, or integrated, crime prevention policy, therefore, rejects the traditional paradigm according to which prevention is viewed only in terms of punishment. Consequently, a new notion of prevention, related to the formation of civic awareness and legal order, has to be construed. This new concept would make preventive actions function mainly in terms of the values that are embedded in the criminal law. That is, it emphasizes legally protected values, which are the basic principles of society (that is, those values which were incorporated in the law and which support beliefs held at a particular place and time).

For a better understanding of this, one should recall that legal norms are either obligations, permissions, or prohibitions, sanctioned as such by the law, and that criminal law imposes punishments on violators. However, the punishment for a given behavior must itself be the subject of a value judgment or appraisal, which is what "determines" that the contrary-to-law behavior has a negative value toward society and is to be condemned.

Thus, crime prevention policy should take as its basis the legal system's actually protected values. Human behavior is oriented by a series of values or preferences which carry with them the possibility of going from what is bad to what is good or better. It is this possibility that makes sense of the efforts of individuals, groups, or nations to overcome ignorance, poverty, social injustice, and violence. This very possibility of improvement is what justifies a preventive policy, whose aim is basically to ensure social cohesion (that is, a state in which criminal behavior would be neutralized).

Prevention should be placed somewhere between ethics and law. Accordingly, prevention strategies should be directed toward the reinforcement and protection of socially accepted values, both personal and institutional. For this reason, instead of emphasizing the punishment of criminal behavior, prevention policies should gravitate around the values that constitute the foundation of the criminal law. The latter should be conceived of as an axiological structure, which responds to the interests and needs of harmonious ways of living (Madrazo and Beller, 1997).

It is not possible to live as a human being without making value judgments. People accept certain things and reject others, either privately or publicly. From a subjective standpoint, values are understood as a preference, since valuing is evaluating. Becoming aware of a system of values is a process that forms a person's evaluating criteria from a very early age. Thus, long-term crime prevention is unavoidably linked to education.

TO PREVENT IS TO EDUCATE

Comte expressed the axiom that to know is to forecast. This also is applicable to crime and drug dependency. Information is essential to foresee preventive actions. The National Addiction Poll, carried out in 1993, revealed that 43 percent of drug users are between 12 and 15 years old. Another poll, which helped analyze national drug trends, made it evident that the age of onset of drug use fluctuates between 10 and 19 years. Likewise, it highlights the fact that cocaine consumption, as an initiating drug, doubled from 2.2 to 4.4 percent within just two years (from 1993 to 1995). This information shows that the drug dependency phenomenon starts at younger and younger ages. At this stage, however, society certainly can speak of a lack of information.

A policy of prevention should not overlook the importance of the early stages in the formation of human personality. This implies that prevention strategies have to be supported to a greater degree by the values on which the structure and functioning of Mexican families stand. In Mexico, the traditional organization of the family, both in a narrow and in a broad sense, is the most important component of the individual's socialization. That is why the Mexican family is the most appropriate medium for propagating the decisive values that give support and stability to the society. Thus, the office of the Attorney General is currently working on a series of agreements with parents associations to spread information and apply strategies that would prevent children and teenagers from using drugs and becoming addicts.

On the other hand, the educational process implies a pedagogy based on the idea (originally expressed by Plato) that virtue cannot be taught. This does not mean that a person is unable to become aware of moral duties with respect to legal order. Ethical abilities and aptitudes are more difficult to foster, in contrast with the rather easy assimilation of abstract knowledge pertaining to the subject's cognitive areas. However, the construction of a moral and axiological personality is feasible and easier in an environment in which the individual's freedom and autonomy is respected.

In Mexico, the educational system aims at developing and applying a pedagogy that, more than merely conveying knowledge, helps the pupil acquire by herself those cultural resources that are essential to the formation of value judgments. Knowledge is necessary but is not enough to prevent criminal behavior or behaviors, which tend to promote drug consumption. That is why prevention is concerned with producing attitudes and personal convictions related to specific lines of behavior, regardless of their being criminal or inducing illicit drug consumption. The office of the

Attorney General of the Republic works jointly with the Ministry of Public Education in order to design and implement permanent campaigns of moral awareness as an essential element of civil culture.

Advertising campaigns do have an impact as preventive elements. However, it is well known that these campaigns can be useful only when they are aimed at specific social groups and when they are provided with specific, concrete goals. Therefore, advertising campaigns must first identify the target public and must evaluate the expected results. Thus, the purpose is not only to select messages that will supposedly exert a positive influence on individuals of a certain age (say, youngsters), but also to consider the cultural background against which these messages can be implemented and employed.

Most essential is that these publicity campaigns be based on clear-cut and well-established concepts, linked to positive values. The axiological basis consists of the fact that what is positive necessarily requires its opposite—that is, what is negative. Behavior that improves the individual and the collective life of society is judged as positive and, therefore, is reinforced. Thus, for instance, the office of the Attorney General fosters what is called "sporting-recreational activities," through which individuals who have succeeded in sports or the arts are stimulated. Obviously, the elements of success are not magical or fortuitous. Rather, they identify themselves with things like the will to face and to overcome difficulties and adverse circumstances. Values like these must be introduced and emphasized in all contexts. In general, the right attitudes are incompatible with violence or with the use and consumption of drugs.

In many ways, seemingly intangible factors that affect our behavior—however indirectly—can have the greatest impact on crime prevention and drug consumption. For example, the declaration of clearly defined rules (which are actually obeyed) by the whole family or by all students at school is a good means of establishing cooperative links that progressively foster awareness of values. Dialogue, instead of authoritarian imposition, the development of a sense of confidence instead of reproach and recrimination, and the respect of the others' dignity, as opposed to exclusion and emotional blackmail, are practices that should be promoted at both home and school. Taking them as a starting point can effectively prevent unlawful behavior.

On the other hand, education with respect to the law (as well as on the use of drugs) should not be approached simplistically. Ethical guidelines suggest that the subjects have to be dealt with from different perspectives at the same time. This is to say that the involved activities have to do with a variety of subjects and should be practiced at different stages of the educational process. The formation of values results in the formation of the person subject to the rule of law, of the citizen, of the moral human being.

THE BASIS OF SOCIAL HARMONY: VALUES

The anthropological and philosophical principle behind crime prevention activities is that people, as a set of individuals, always have interests and that these interests are the objective factors which are decisive for their well-being. The greater the satisfaction of these interests, the higher the level of well-being. Some specific interests get cultural and legal recognition, since they are expressed as juridically protected values. Criminal law contains a whole catalog of values referring to behavior. These values aim at preserving and protecting those interests which are the basis of social peace and which make personal development, as well as material and spiritual progress, possible.

It is precisely for these reasons that all crime prevention efforts must be directed at the protection of those juridically regulated values, which in the last analysis are fundamental and necessary for individual and social well-being. Prevention policies must go well beyond its purely punitive component and promote the individual's and the group's improvement. After all, crime prevention is nothing but a set of actions directed at the promotion of stability and social development.

For example, drug trafficking is typically a crime against society's health. To that extent, it puts at risk not only collective health, but other underlying judicial goods as well. Individual freedom is endangered since the deprivation syndrome or abstinence causes the capacity to choose goals freely to deteriorate or even to end. Fiscal interests are threatened since drug trafficking is at the margin of official controls. Public security suffers, since drugs contribute to an increase in criminality. Drug consumption hurts personal and social morals by promoting uses and habits contrary to family and community stability. The national economy suffers, because drug traffic is an activity that prompts laundering money, tax evading, and so on. Finally, national security itself is threatened, because drug dealers' organizations maintain links with international groups devoted to weapon smuggling and trafficking and to international prostitution.

A crime prevention policy that concentrates on the strengthening of culturally accepted and juridically protected values must therefore be implemented within the framework of a social and economic policy planned and set in motion by the state, among other things, because social and economic circumstances are very often the conditioning (though not deciding) factor for higher levels of criminal action.

No preventive action can be effective or long-lasting unless it has the full support of the people. This means that prevention campaigns must first take into account the goals people attempt to reach. It has to consider their characteristics, such as educational level, social status, and specific cultural expressions. On the other hand, those campaigns must convey and teach a positive message, a message which would encourage lawfulness, confidence in state institutions, peaceful coexistence, and harmony in social relations. The point of the above is to generate enthusiastic collective action in support of the efforts being made by the authorities, because there cannot be any real prevention without social collaboration.

Crime prevention requires a very specific analysis of crime zones and of data produced by the best possible knowledge of social, economic, and cultural circumstances associated with criminal behavior. Any problem shows its complexity if it cannot be solved by a single perspective approach and if the treatment requires the simultaneous presence of several scientific points of view. The phenomenon of crime is increasingly complex. Today, crimes are planned more carefully and carried out by means of more sophisticated technology than they were in the past. Crime prevention efforts therefore should be supported by specialists from diverse areas of knowledge, and the institutions charged with the administration of justice must be made responsible for the organization of relevant knowledge. Prevention efforts need criminologists, specialists in communication, lawyers, medical practitioners, social workers, anthropologists, and psychologists. It is only with their joint efforts that adequate crime prevention strategies can be planned, developed, applied, evaluated, and improved.

Specific Programs for Crime Prevention

Crime prevention policy should have as its goals the analysis of the risk factors involved in criminal activities, the elements that are supposed to protect the individual's rights, and the implementation of the results of criminological research. Since November 1977, the office of the Republic's Attorney General gave six courses to train promoters working on drug addiction prevention. These courses, sponsored by UNICEF and the National System for Family Integration, took place in the states of Zacatecas, Quintana Roo, Durango, Michoacán, Guanajuato, and Jalisco. These training courses were attended by 950 persons who had wide recognition in their own communities. The courses were evaluated thoroughly by institutions at the federal, state, and municipal levels.

It is clear that crime prevention must point to specific actions that are not limited by the spread of information as to the consequences of illicit behavior or the consumption of forbidden substances. Thus, since July 1977, in five states of the Republic (Sinaloa, Jalisco, Chihuahua, Baja California, and Coahuila), a preventive project labeled Sport Recreation Units has developed. The main goal of this project is to acquire properties and to prepare them for sports practice in areas where high levels of criminal activity or drug use have been observed, and then to hand them over to local inhabitants. This project is being carried out by the office of the General Attorney together with the National Council Against Drug Addiction (of the Health Ministry), the National Sport Council (of the Ministry of Public Education), and the National System for Family Integration.

Enjoying sports activities is highly recommended for the reduction of criminal behavior, including the use of drugs or psychotropic substances. Nevertheless, sports by themselves are not fully effective unless they are practiced in conjunction with other factors. The point of practicing some sport is that it implicitly promotes values like discipline, effort, perseverance, caring for oneself, and so on. These are values that are incompatible with crime and drug addiction. The Sport Recreation Units represent an attempt to encourage precisely these values, explaining to the participants in the program, through talks and debates, why they are important and why they should be respected.

These units have been created and promoted in order to provide services for young people, inviting them to carry out different activities within them. There are three main goals. The first is *prevention*, both of criminal activity and in the area of health (individual and collective). The second is the *sport activity* itself, to strengthen each child's personality and help him to acquire and develop values that are intrinsically opposed to those involved in the use and consumption of drugs. Finally, *recreation* is understood as invaluable for the kind of entertainment that encourages the community's traditional values, cultural identity, and optimal use of free time by people of both sexes.

From the point of view of this chapter, the war against corruption should begin with the process of molding young people by making them accept and respect civic values. This is a task very similar to what is being done by the Ministry of Public Education. There are, however, no programs that explicitly face the issues entailed by the war against corruption. Organized crime has only very recently been detected in Mexico and is no doubt a subject of great complexity. It is important to recall that the federal Law on the Combat of Organized Crime was enacted as recently as November 1996.

Organized crime is affecting an ever larger sector of the population, which makes it imperative to devise new intervention strategies. The office of the Republic's Attorney General is currently embarked on an effort to inform the people of Mexico regarding the content and consequences of the new law. Simultaneously, the Ministry of Finances is preparing a manual to help banking institutions protect themselves from money laundering activities.

In summary, an adequate crime prevention policy can be implemented only as long as it is enshrined in a wide-ranging cultural, conceptual context in which differing and varied points of view are acknowledged. Such policy should be carried out within the framework of democratic structures. If it is true that prevention strategies are basically, though not exclusively, based on moral education, it also is necessary to admit that value differences or clashes may incorporate ideological, technical, political, and religious problems. Well-educated people, however, should be in a position to solve their value conflicts in a peaceful manner. On the other hand, nothing should prevent joint institutional and social action from being carried out.

Instilling moral values is one potential solution in the common effort to curb antisocial behavior, but it should not be sought as a remedy for everything. There is still much work, much planning and evaluating to be done. The problems are of such magnitude that all efforts must be unified and intensified. Certainly, the goal makes the effort worthwhile.

The present trend, which holds much hope for the future, is to turn all crime prosecuting institutions into agencies whose most important objective is to prevent antisocial behavior. Education is and will be an essential and irreplaceable instrument in the achievement of this new reality.

SUMMARY

The actions of procuring justice are carried out under an integrated or systemic approach, which means excluding isolated actions and ignoring the cultural, economic, and political contexts in which they are realized. Crime prevention programs are a subsystem of justice procurement and administration. The prevention of antisocial behavior is sustained by the idea that government institutions should contribute to building a civic consciousness which would express itself in respect for the law and for human rights. Essentially, preventive programs are programs of social pedagogy that look to assist the judicial system and the values it embodies. Therefore, prevention policies seek to develop a civic consciousness based in each person's ethical autonomy. To prevent is to educate.

No prevention program will be available if it does not include, in an integral way, the participation of the population. This means that the combined participation of both individuals and state institutions must be encouraged.

CHAPTER 15

CRIME PREVENTION IN CHINA: A COMMUNITY POLICING APPROACH[1]

Kam C. Wong

Our public security work. . . [should] not . . . have matters monopolized by the professional state agencies. It is to be handled by the mass . . . The mass line principle . . . is to transform public security work [into] the work of the whole people.

Minister of Public Security
Luo Ruixing (1994)

It is only a slight exaggeration to say that if American crime prevention is a device by which citizens assist the police, in China it is seen as a method by which police provide back-up services for citizens.

Professor Dorothy Bracey (1984)

Crime prevention through community policing is a worldwide growth industry (Skolnick and Bayley, 1988). During the 1980s, England (Schaffer, 1956; Friedmann, 1992:172; Weatheritt, 1986), Singapore (Quah and Quah, 1987), and Canada (Friedmann, 1992:183) introduced various forms of community policing to fight crime. In the United States, community policing originated as a "quiet revolution" seeking recognition during the 1970s (Sherman, 1974; Kelling, 1988) and has since become a tour de force to be reckoned with in the 1990s (Bayley and Sheering, 1997). The community policing movement, however, has not been universally well received. Whereas the Japanese police have had great success with their *koban* community centers (Bayley, 1976:15–632), the Australian police have to work hard to keep their Neighborhood Watch alive (Bayley, nd; Soloff, nd), and the Hong Kong police have had difficulties in engaging the public in tending to community problems (Chiu, 1996:60–61).

Turning to China, social control has always been provided for by the local communities, grassroots organizations, and intimate associates. Historically, much of social life was regulated by grassroots and indigenous social institutions (Sprenkel, 1962), such as family (Wilbur, 1978:113–175), clan (Liu, 1959), village (Smith, 1899; Hu, 1984; Yang, 1947), guild (Chan, 1975:28–40), and voluntary associations (Wong, 1971:62–73). More recently, the People's Republic of China mobilized and empowered the mass to take control of their own community (Luo, 1994:57; Bennett, 1977:121–139).

The Chinese experience in community policing, until recently, has been a positive one, as evidenced by community solidarity (Topley, 1967:9–44), communal activism (Li and You, 1994:124), and a low crime rate (Li and You, 1994:40). This compares favorably with the U.S. experience (Skogan, 1990), which is characterized by community fragmentation (Reiss, 1983:43–58; Rosen, 1990:9–12), apathy (Wilson and Kelling, 1982:29–38), dependency (Gurwit, 1933:33–39), clienthood (Peak and Glensor, 1996:46-47; Walker, 1992:189–190), and a considerable fear of crime (Skogan and Maxfield, 1981:121). The contrast between United States and Chinese community policing, in terms of philosophy and practice, is great and deserves further in-depth investigation (Pepinsky, 1973, 1975; Braithwaite, 1989) and critical examination (Birkbeck, 1993).

This chapter is an investigation into rural community policing in the People's Republic of China (PRC)—its philosophy, law, practice, and issues. The first part presents the focus of and justification for this research into PRC rural community policing. We then proceed to a discussion of the philosophy of Chinese community policing, past as well as present. This is followed by a brief overview of the laws of community policing in China. PRC rural community policing in practice is then described and some of the salient issues and emerging problems associated with rural community policing are identified. The last part of the chapter discusses the lessons provided by this investigation into PRC community policing.

A DEARTH OF RESEARCH

There is no basic English text on present-day policing in China. There are a few dated introductory texts on the PRC criminal justice system (Cohen, 1968; Brady, 1982; Leng and Chiu, 1985) and research monographs into PRC social control methods and measures (Troyer et al., 1989; Wilson et al., 1977; Dutton, 1992) that touch upon policing tangentially. More recently, several articles were written on various aspects of policing in the PRC (Johnson, 1984; Ward, 1985; Ward, 1985a; Patterson, 1988; Wong, 1994, 1996, 1996a, 1997; 1997a, 1998, 1998a). Very few studies on crime prevention and community policing in China have been made. The rare exception is the investigation by Zhang and his colleagues (1996) into community crime prevention in Tinajian City. This research gap resulted in part from a general lack of scholarly interest, exacerbated by research difficulties posed by language, culture, and secrecy.

Students of PRC rural community policing have to look elsewhere for information and inspiration. Some often-neglected sources in the literature are the historical accounts of communal rural social control at the family, clan, village, community, and neighborhood level (Liu, 1959; Hsiao, 1960; Madsen, 1984; Wen, 1971). In as much as Chinese rural communities, until very recently, were relatively closed and stable sociopolitical units, studying the past helps inform the present. The other valuable sources are the few richly chronographed ethnographic studies of PRC village social life. They provide rare empirical data on the day-to-day operations of rural-communal social control in China (Hinton 1966; Parish and Whyte 1978; Huang 1986). Finally, PRC criminologists and police scholars have provided some authoritative documentary accounts on law and policy bearing on the rural social control system (Yanli, 1994:229–234; Li and You, 1994:111–118).

This cursory review of the literature shows that there is currently no comprehensive study of rural community policing in the PRC. This chapter is a first attempt to fill that gap.

PHILOSOPHY OF COMMUNITY POLICING IN MODERN CHINA (PRC)

In defining the police-community role in fighting crime and keeping order, the PRC has taken the approach that

> Our public security work . . . [should] not . . . have matters monopolized by the professional state agencies. It is to be handled by the mass. . . . The mass line principle . . . is to transform public security work into the work of the whole people. (Luo, 1994:57).

To this most interesting community policing philosophy we now turn.

The Burden of the Ubiquitous Past

Historically, social control in China was decentralized and organized around natural communal and intimate groups, such as the family and clan (Hsiao, 1960:261–371; Troyer, 1989:16–24; Dutton, 1992), with governmental endorsement and support (Wong, 1998a). Local social control was institutionalized. The emperor ruled the state by and through his officials, who in turn governed the people by and through the heads of families and community leaders (Wen, 1971; Chang, 1955). Such decentralized, grassroots social control was informed by Confucian teachings:

> Wishing to govern well in their states, they would first regulate their families. Wishing to regulate their families, they would first cultivate their persons. Wishing to cultivate their persons, they would first rectify their minds. Wishing to rectify their minds, they would first seek sincerity in their knowledge. Wishing for sincerity in their thoughts, they would first extend their knowledge (Bary et al., 1960:115).

Thus, historically and traditionally, functional social control in China was supplied informally and outside of the courts. This resulted from a deliberate state policy to build upon existing natural communal structure (Dutton, 1992:84–85), ancient cultural habits (Williams, 1883:507; Chu, 1962), and deeply rooted customary practices (Liu, 1959:56). Hence, although in theory the local magistrates were supposed to be in total control on all matters large and small in rural China, in actual practice, broad police powers were conceded to the local community to be exercised by the family (Liu, 1959:chap. 2; Wittfogel, 1957:50).

PRC crime control policy and practice are very much influenced by historical Chinese (Confucian) thought. More particularly, Chinese traditional thought on social regulation and crime control were informed by the following premises:

1. Crime control is a local, indigenous and, above all, family affair (Bary et al., 1960:115; Chu, 1962:chap. I; Wu, 1995:437–510; compare with Wilson and Herrnstein, 1985:213–244).
2. Prevention is the first step in crime control. It addresses the early symptoms and manifestations of personal problems (Feng, 1994:67–76; compare with Wilson and Kelling, 1982:29–38). Collaterally, successful crime control looks for the root causes, and not the outward symptoms of crime, such as the moral degeneration of the individual, the cultural pollution of the people, and criminogenic social conditions of the community (Bary et al., 1960:115).
3. To be effective, crime prevention must be a multifaceted, comprehensive, and integrated enterprise, involving the individual, family, clan, neighbor, community, and, finally, the state (Bary et al., 1960:115; compare with Peak and Glensor, 1996:88–92).
4. To succeed, crime control and prevention measures should be variegated. Confucius observed that the best way to regulate people is to "inspire them with justice, correct them with administration, guide them with rites, keep them straight with honesty, appeal to them with benevolence, reward them with benefits, and persuade them to follow" (Lee, 1988:61; compare with Peak and Glensor, 1996:411–414). More simply put, crime control can be best achieved through moral education as supplemented by immediate, severe, and speedy punishment (Liu and Yang, 1984:56–64).

Consistent with the above Confucian ideas and ideals, crime prevention and social control in traditional China was realized through indigenous groups starting with the family, which provided the education and discipline for character building, the neighbors, who provided the supervision and sanction against deviance, and the community, which set the moral tone and customary norms to guide conduct. Finally, the state acted as the social control agency of last resort in providing legal punishment for crime, and economic maintenance and social welfare to anticipate civil disorders.

All told, the Chinese came up with a broad basis of control (Wittfogel, 1957). Control existed at macro-, intermediate, and microlevels (Gibbs,

TABLE 15-1 COMPARISON OF TRADITIONAL SOCIAL CONTROL PHILOSOPHIES: EAST AND WEST

	East (China)	West (United States)
Justification	Reformation (of offender) Restoration (of social relationship) Reintegration (for communal harmony)	Retribution (to victim/society) Deterrence (to individual/society) Rehabilitation (of individual)
Subject	Personal character—internal thought	Social conduct—external behavior
Basis	Moral wrong	Legal wrong
Method	Education—to reform	Punishment—to deter
Strategy	Root of the problem Preventive—proactive	Manifestation of the problem Remedial—reactive
Site	Collective	Individual
Sources	Multiple layers—individual, family, neighbor, clan, state Multiple focus—psychological, physical, social, economical, legal, political, cultural	Unitary system—judicial, legal
Nature	Informal, social	Formal, legal
Time	Proactive	Reactive
Assumption of the controlled	Affective, social	Rational, autonomous

1982:9–11) that included the internalization of norms (by the individual), socialization and disciplinary regimens (by the family), the setting up of custom and accountability systems (in the community), removal of criminogenic conditions (by the administration), and a defining of moral and social boundaries (by the state).

The Influence of an All-Consuming Present

The philosophy of PRC policing, conceptually and operationally, is determined by the Chinese Communist Party's political ideology. The PRC leadership from Mao to Deng to Jiang advocated that the people, or "mass," are masters of their own destiny (Yanli, 1994:18–19). In this regard, Dong Biwu stated emphatically:

> The power of the border government came from the mass (*qunzhong*). . . . The government has to listen to the mass, adopt the mass's viewpoints, understand the mass's life, protect the mass's interest. And that is not enough. It must see to it that the mass has courage to criticize the government, supervise its work, and replace unsatisfactory workers. . . . Only through this can the mass feel that the government is a tool in their own hand and the government really their own government (Peng, 1991 : 104).

In the early days of the PRC, Mao practiced the "mass line" doctrine by using individual voluntarism and mass mobilization in conducting the people's business and to move the country ahead (Zhang and Han, 1987: 75–76, 99). Mass struggle and self-criticism, not legal norms and judicial process, were used in ordering society (Brady, 1982).

In practice, following the mass line meant that Party should work and live among the people in order to understand their problems. They should also listen carefully to people's concerns in order to formulate solutions to problems. Ultimately, the people's own "ideas are preserved . . . and carried through" (Brady, 1982: 69). "In all the practical work of our Party, all correct leadership is necessarily from the masses to the masses" (Mao, 1967: 226–227). More recently, under the leadership of Peng Zhen and later Qiao Qi, "power to the people" was realized through more structured means, that is the constitutionalization of the political process and legalization of democratic rights.

When the idea of the mass line was applied to justice administration, "popular justice," "informal justice," and "societal justice" were preferred over those which were formal, jural, or legal. Informal justice models involve the mass in the settling of disputes, such as through mediation. Popular justice engages the mass in the dispensation of justice, such as by public denunciation (Leng and Chiu, 1985:11). For example, during the earlier years of the PRC (between 1950 and 1951), the mass was allowed to dispense with revolutionary justice during the land reform, "three-anti" (*san-fan*) and "five-anti" (*wu-fan*) movements (Editorial, 1979:25). In later years and in judicial work, court trial proceedings were integrated with mass debates, effectively bringing the courts to the people (Cohen, 1968:11). Mass trials were held in public, not to ensure a fair trial but to educate the people, raise their consciousness, and empower them.

In terms of policing, the mass line formed the basis of "people's policing," whereby the people were responsible for policing themselves. In the context of fighting against political crimes (and later all social crimes), the people's participation was deemed indispensable in winning the "people's war" (Sun, 1977:1–4). Thus, in the suppression of counterrevolutionaries, the police were supposed to understand the people, trust the people, mobilize the people, and rely on the people (Luo, 1994:82). The police played a supplemental (not dominant) and facilitating (not instigating) role. In all, the people were considered the lifeblood and backbone of the police. The people and public security could be said to be "coproducers" (Rosenbaum et al., 1991:96–130) of revolutionary order and justice, with the people being the principal partner (Luo, 1994:93).

Practically and operationally, people's policing meant that the police must see things from the people's perspective, seek their support, and be amenable to their supervision (Luo, 1994:189). If the police were detached and isolated from the people, they would not be effective in ferreting out local problems and addressing domestic concerns (Luo, 1994:213; Jiang, 1996:9). Hence, one of the more serious "mistakes" that can be committed by the police is having an erroneous work style—that is, being alienated from the mass through subjective idealism, bureaucratism, and commandism (Luo, 1994:179, 189).

The people must be involved and actively engaged in fighting the "people's war" against crime for many different reasons. First, the people have the right to participate in their own governance. This dictum is akin to the idea and ideal of localism in the United States (Briffault, 1990:1–115; Briffault, 1990a:346–456) wherein all the powers of the central government come from the people. The spirit and essence of localism is best captured by the United States Supreme Court opinion in *Avery v. Midland County:* "Legislators enact many laws but do not attempt to reach those countless matters of local concerns necessarily left wholly or partly to those who govern at the local level" (390 US 474, 481 [1961]).

Second, the people have the responsibility as citizens to fight crime. In the PRC, people's rights and responsibilities are complementary. Article 33 of the PRC Constitution (1982) provides that "citizens enjoy rights guaranteed by the Constitution and law but they must also fulfill their constitutional and legal responsibility." This is similar to the notion of "communitarianism" in the United States, which declares that "the whole community needs to take responsibility for itself. People need to actively participate, not just give their opinions . . . but instead give time, energy, and money" (Gurwit, 1933:33–39).

Third, the people are in the best position to see that "people's justice" is done, including making decisions regarding who to police, what to police, and how to police. Mao supplied the rationale for "people's policing" in his "Report on an Investigation of the Peasant Movement in Hunan":

> The peasants are clear-sighted. Who is bad and who is not quite vicious, who deserves severe punishment and who deserves to be let off lightly—the peasants keep clear accounts and very seldom has the punishment exceeded the crime (1967:28).

This is similar to the idea in the United States that the community notion of order and justice prevails over the rule of law (Wilson, 1968:287).

Fourth, the people are deemed to be more motivated, thus more vigilant, as an oppressed class, to detect the counterrevolutionaries (Luo, 1994:57). This reflects the notion that citizens of a state, as with employees of an organization, naturally seek responsibility if they are allowed to "own" a problem. "The average human being learns, under proper conditions, not only to accept but to seek responsibility" (McGregor, 1960:48).

Fifth, the people are in the best position, being more able, efficient, and effective in conducting their own business. Criminals and counterrevolutionaries live in the mass. They cannot long survive within the mass without being exposed by them (Luo, 1994:292). This mirrors the notion in the United States that the public is the best source of intelligence for the police (Sparrow, 1993:4).

Finally, it is unrealistic to expect the police to provide the optimal level of security without assistance from the people. The police are not omnipresent nor are they omnipotent. This is especially true in the sparsely populated areas, such as border and rural areas (Luo, 1994:292). It is most unlikely that the police could be informed of all the illegal activities taking place unless they are assisted by the people (Luo, 1994:347–352).

Put into practice, "people's policing" in the PRC requires that the police work closely and intimately with the mass (Luo, 1994:315). This means (Luo 1994: 309–311) that the

1. Police and the people are inseparable
2. Police should work with the best interest of the people in mind
3. Police should involve the people in preventing, detecting, and controlling criminal activities
4. Police should educate the people on the state's crime control policy
5. Police should engage the people in the execution of state law and policy
6. Police should welcome the people in supervising police actions
7. Police should avoid being isolated from the mass

However, the people are sometimes wrongheaded, ignorant, apathetic, and not well organized (Hsiao, 1960:264). The police, as political leaders, must mobilize the people, raise their political consciousness, and try to organize them (Luo, 1994:72).

LEGAL BASIS OF COMMUNITY POLICING IN THE PRC

Article 2 of the PRC Constitution of 1982 provides that "all powers of the PRC belong to the people. . . . The people in accordance with the law, through various methods and forms, manage state affairs . . . [and] social affairs." Specifically, Article 111 provides that the village communities should be self-governing:

> The residents' committees and villagers' committees . . . are mass organizations of self-management at the grassroots level. . . . The residents' and villagers' committees establish subcommittees for people's mediation, public security, public health, and other matters in order to manage public affairs in their area, mediate civil disputes, help maintain public order, and convey residents' opinions and demands to the people's government.

The Organic Law of PRC Villagers Committee (Provisional) empowers the rural village residents to govern themselves and manage their own affairs (Article 1) (Police Laws, 1990:2167–2169). The "villagers' committee is the villagers' self-regulating, self-educating, self-servicing local mass self-governing organization" (Article 2). The functions of the villagers' committee include "resolving civil disputes, assisting in maintaining social order, reflecting villagers opinions, and making demands and suggestions" (Article 2).

On June 27, 1952, the Government Council promulgated the Provisional Act of the PRC for the Organization of Security Defense Committee (SDC Act) (Police Laws, 1993:1401–1404). The SDC Act provided for the establishment of the Security Defense Committee (SDC) in local communities to secure law and order. As conceived, the SDC was instrumental in "preventing treason, espionage, theft, and arson, in liquidating counterrevolutionary activities, and in defending state and public security" (Article 1).

Since 1978 and with the opening of China, the role and functions of the SDC have been redefined. The SDC is now mainly concerned with providing for public order through crime prevention, such as community crime watch, mutual surveillance, and order maintenance, including mediation of disputes and supervision of offenders (Police Laws, 1993:1401). More recently, the SDC has shifted all its attention and resources to fighting crime. In so doing, it has all but given up its political mission to attack political enemies at the behest of the state (Police Laws, 1993:1401).

ACTUAL PRACTICE OF COMMUNITY POLICING IN THE PRC

How Community Policing Functions

The SDCs are delegated with three policing functions. First, one of the most important functions of the SDC is to resolve incipient conflicts and disputes through mediation (*tiaojie*) before they become unmanageable. Conflicts and disputes are resolved by the SDC-Mediation Committee with reference to law, regulations, custom, and village agreement. The village people prefer mediation because it is informal, inexpensive, and effective. The village people also do not like going to the officials—bureaucrats and outsiders—to resolve local and personal problems. Lastly, a successful mediation resolves a dispute without disrupting existing relationships.

Second, the SDC is responsible for organizing the local security defense. For example, in May 1995, the local security system of Kedong county, Shangxing province, Jianshe village consisted of four specialized, full-time, security defense teams directed by a captain, who was also an

SDC member. The security defense personnel were paid $2,000 a year and funded by a household tax of $1.80 per month per household. Each security defense agent had to sign an undertaking, secured by a $4,000 guarantee or property pledge, to carry out his responsibility. He had to repay the village for any property and animal lost during his shift, specifically 80 percent of the value of any stolen vehicle and 50 percent of the value of any stolen grains. The security responsibility system placed the crime prevention squarely in the hands of the local SDC and its security team agents. This proved to be effective in preventing crimes and reducing losses.

Third, the SDC is responsible for the proper maintenance of household supervision within a village. A responsible person in each village is designated as an official household agent in charge of household registration matters. The person works closely with the household police officer at the police post. The SDC keeps four kinds of household records. One is a book listing all persons without a registered household and any registered households without some or all of its registered residents. The second is a chart containing the name, residence, and household address of all the residents in the area. In addition, a photo album is kept of all the youth and able-bodied people in the area between 16 and 50 years of age. Finally, a map is kept showing the distribution of households, including the location of SDC directors and residents of "focus group people." The household registration system allows the SDC and police to keep track of the coming and going of suspected criminals from without and disruptive activities of troublemakers from within (Dutton, 1992).

Community Policing in Action: A Case Study

How community policing works in practice can be best illustrated with a typical case supplied by the Minister of Public Security for propaganda and educational purposes (Wen and Chen, 1997:25–27). In 1993, a 28-year-old public security officer, Mr. Sui, was transferred to Lun Yang Zhen, a rural town with 50,000 people, as the head of the local police post. The Lun Yang Zhen was a newly established rural town 46 kilometers from the nearest city. It had no paved road. The community suffered from a lack of established policing authority since it came under nobody's jurisdiction. It had a serious public order problem for many years.

When Officer Sui arrived, there was no office, car, or communication equipment. Although he had a staff of three police subordinate officers, Sui was for all intent and purposes the only public security officer in town. The first thing he did was to recruit and form a 10-person joint-security defense team. He also set up a village protection unit in each village. He spent another month organizing SDCs in the 27 administrative villages under his charge. He built 277 guard posts and watch houses for his SDCs. He divided the policing area into five police districts and four police beats. He assigned policing and order maintenance duties to each of his administrative village Party secretaries, SDC chiefs, and SDC members. He established a security "responsibility system" and started night patrols in affected areas.

Upon arrival, Officer Sui personally visited each of the villagers. With his selfless devotion to duty and compassion for people, he was able to earn the residents' trust and respect. He established his authority by steadfastly standing firm against any political influence from above and economic corruption from below. He demonstrated leadership by personally supervising the SDC's maintenance of law and order. He checked the household registers religiously to sort out criminal elements and relentlessly conducted criminal investigations to reduce crime. He patiently mediated disputes to prevent the minor conflicts from escalating into major crimes.

This case study makes it clear that in order to be effective, the police must work hand in glove with the people, starting with an understanding of their problems and earning their trust. In fact, the imagery often used to describe the relationship between the police and the people is that of fish (the police) and water (the people): Fish cannot long survive out of the water, just as the police cannot effectively function without the people.

COMMUNITY POLICING IN CHINA: ISSUES AND PROBLEMS

The SDC—Changes in Roles, Functions, and Methods

As originally conceived under the Provisional Act of the PRC for the Organization of Security Defense Committee passed in 1954, the SDC was a political institution. It was established as an indigenous mass organization to mobilize the people to eliminate all classes of enemies—traitors, spies, counterrevolutionaries, criminals, and social misfits—and to build a communist revolutionary order (a proletarian dictatorship). After 1978, the SDC was reoriented to deal with social security and public order issues and concerns.

On January 24, 1980, the Ministry of Public Security promulgated the Notice Regarding Continuing to Implement the Provisional Organic Defense Security Committee Regulations (SDC Notice) (Zhongguo Gongan Bake Quanshu Editorial Committee, 1989:226). The SDC Notice provided for the continued application of the 1954 SDC act because it changed the primary role of the SDC from one of securing revolutionary justice to that of providing for law and order. Instead of seeking to raise the people's class consciousness through mass campaigns, it now attempts to educate the people in the spirit and letter of the law. Rather than mobilize the people to struggle against class enemies, it now organizes the people to provide for mutual defense against criminals. In all, during the 1990s, the SDC fought crime and maintained order with the rule of law and by means of established legal processes. It no longer imposed political domination over class enemies with the use of mass mobilization and public struggle.

Rule of Law versus Norms of Community

The villagers' committee, and by extension the SDC, is intended to be a self-governing (self-managing, self-educating, self-servicing, self-ordering) grassroots institution (Li and You, 1994:4). It operates on the principle of democratic self-governance (Li and You, 1994:113). However, in theory as well as in practice, it is not supposed to be totally independent and autonomous, that is, functioning freely without any limitations imposed by the Party, state, or law (Li and You, 1994:114). In reality, the SDC's role and functions, duty and responsibility, power and authority are very much affected by communist party policy, state law, and administrative directives. There are structural tensions and inherent conflicts in having the SDC function as an autonomous self-governing unit and in subjecting it to institutional control (legal, political, administrative) of one form or another. The PRC political authority is keenly aware of such tensions (Li and You, 1994:4).

In all, three kinds of structural tensions are observed (Li and You, 1994:129). The first involves Communist Party policy versus community autonomy. Political imposition is justified because a community's self-rule is allowed solely to achieve communist ideals and programs, and a community's self-rule can be properly exercised only by and through the leadership of the communist party (Li and You, 1994:113). The second is central administrative direction by the government versus local democratic rule. Administrative imposition is justified because a community's self-rule must be structured to implement state policy and not be in conflict with state policy and objectives (Li and You, 1994:113). The third is state law and regulation versus communal customs and norms (Gao, 1994:11–16). Legal imposition is justified because a community's right to self-rule must be exercised within the confines of the constitution as empowered by law.

Article 2 of PRC Constitution provides that "all powers of the PRC belong to the people. . . . The people in accordance with the law, through various methods and forms, manage state . . . social affairs." In sum, a community's self-rule is to be exercised within a well-defined "top-down" framework (politicized, legalized, rationalized, and bureaucratized) to be justified on grounds of legitimacy, utility, and shared interests (Li and You, 1994:129).

External impositions on a community, however legitimate and justified, are often resented as gratuitous interference, if not as unwarranted encroachments on local autonomy. Take the case of legal penetration into the local community (Moser, 1982:174–175). Increasingly, the state has legitimized its penetration into local affairs by and through the law. This is achieved in two ways: (1) by educating the people to the letter, spirit, and rule of the law (Li and You, 1994:114) and (2) by insisting that the local conduct norms and rules should incorporate state substantive law and process (Li and You, 1994:132). This raises squarely the issue of "rule of law" versus "norms of community" (Mastrofski and Greene, 1993:80–102; Bayley, 1988). The conflict arises when a community seeks to protect its interests and the collective welfare at the expense of legally protected individual rights or due process (Wilson, 1968). Consequently, the community sense of justice must give way to legal justice.

On a still-larger compass, the debate on the feasibility of state law versus community norms raises the more general question of whether monolithic legal values, both substantive and procedural, can adequately address diverse local conditions, values, and concerns (Wright, 1980:19–31). More specifically, community norms are, by nature, contingent (for example, on social status, on past conduct, or on balancing utilities), not absolute (based on natural rights and legal entitlements) or rational (Feeley, 1997:119–133), as provided for by law.

Ideological versus Instrumental Community Policing

Leadership philosophy and style has a defining influence on Chinese polity. Mao was an ideologue, Deng a pragmatist, and Jiang a technocrat.

Community policing under Mao was ideological (Wen and Chen, 1997:25–27), subsumed under the broad rubric of the government of, by, and for the people (Mao, 1971:25–27). The government and the people are not separable entities. They are one and the same. In the end, both the people and the police serve the same master—communist ideology. Lei Jingtian, a border area Justice, observed,

> Making justice administration work the mass's own work, the judicial organs the mass's organs, be at one with the mass, listen carefully to their suggestions, respect the fine habits of the mass, fairly and responsibly resolve the mass's problems, organize mass trials without concern with formalities, and reduce litigation of the mass (Yang and Fang, 1987:72).

The implications of this mass-line doctrine, as applied to policing, is clear: The people and not the police are in control ("Making justice administration work the mass's own work."). Nothing is to come between the people and their own aspirations, wants, and realization of self ("organize mass trial without concern with formalities"). In this regard, the police are an extension of the people, of which they are an inseparable part. The police adopt the interest, viewpoint, and method of the people ("listen carefully to their suggestions, respect the fine habits of the mass, fairly and responsibly resolve the mass's problems"). Serving the people is an end, not a means.

Deng spearheaded "pragmatic policing." Jiang followed with "scientific policing" (Wong, 1998). Whether it is pragmatic policing or scientific policing, the result is the same. Policing in China is getting to be more "instrumental" than "ideological" (Feng, 1994:81). In this regard, community policing is instrumental in two senses (Sheng, 1997:29–31). First, local SDC (community policing) is being used increasingly as a policy arm of the communist party and central government working through the local police apparatus. This conforms to traditional communist political wisdom. Second, the police are increasingly using the people to achieve their organizational missions and operational objectives. This action is driven by the need for modernization of the police. The police organization is becoming increasingly more bureaucratized (e.g., division of labor, centralized), professionalized (e.g., scientific crime detection, MBO), and legalized (that is, rule by law) (Xiao, 1997:15).

Dysfunctional Community Policing

Economic reform in China is having a devastating impact on community policing. The impact is being felt in three ways. First, economic reform has led to fundamental social-structural dislocations, resulting in a more criminogenic social environment (Zhang, 1997:4–12). The economic reform has effectively transformed the basic characteristics of the Chinese society, especially the rural communities, from a closed, static, stable, collective, uniform, simple, and tightly controlled society to an open, dynamic, individualistic, diverse, and complex one (Zhang, 1997:4–5). At an individual and psychological level, the economic reform and attendant social transformation has created adjustment problems for individual people (Neal, 1971:103–135). The economic reform has rediscovered in the collective personality an individual identity, characterized by a self-centered, egotistical tendency and a norm-defying profit orientation. The historically suppressed egotistical self and politically subdued profit motive, once liberated by Deng to fight the vanguard battle against the planned economy and collective welfare state, now knows no bound. The preoccupation with an all-embracing self (egotism) as fueled by an insatiable obsession with profit (hedonism) strong enough to be the engine of economic reform, is also not amendable to established communal, informal, and static social control of the traditional kind.

Second, economic reform has weakened existing community policing (social control) structure and methods (Dai, 1997:46–50). Community control in the past was built upon a relatively closed, static, stable, uniform, and simple society in which (community) social controls were strictly enforced. Morality was internalized. Personal character was relatively fixed and immutable. Social, administrative, and political controls were designed to regulate every aspect of a person's value system, thought, and action. This kind of social control is disappearing.

Economic reform has tranformed the once-closed and stable community into one that is open and dynamic. In the process, the static control system of the past proves dysfunctional in containing dynamic deviance. This is most evident in the case of the household registration system. In the past, the key to crime control was through household management, which facilitated mutual surveillance and government control. Effective crime prevention and detection, however, depended on timely and accurate registration of all households. But the registration system is fast breaking down. For example, there is a phenomenon called "*koudai hukou*" (pocket household). The *koudai hukou* is an individual who travels around with a household registration he can readily use (that is, a household readily available "in his own pocket"). People do not register households for a number of reasons. A person might want to avoid state-imposed duties, from those on exceeding the government birth control limit to those involving village duties and levies. Also, local governments and administrative units (like administrative villages) might not register a household in the attempt to show that they have achieved planned economic targets mandated by their political boss (for example, reaching and exceeding "middle class" living standards without unemployment) (Wu and Lu, 1996:23).

Third, economic reform has attenuated, if not completely destroyed, the close and intimate relationship between the mass and the police. As the police become more professionalized and specialized, they become more isolated and alienated from the people they serve. For example, the police are less interested in establishing a rapport with the people and far more interested in catching criminals and fighting crime. In pursuing aggressive crime fighting strategies (for example, indiscriminate fight-crime campaigns and stop-and-search operations), the police alienate local residents whom they are supposed to serve and depend on, and to whom they are accountable. Increasingly, police officers are seen as imposing upon and dictating to the people rather, than engaging and involving the people in beneficial activity. In adopting scientific crime fighting methods, the investigation and detection of crime is made the responsibility of specialized units. This alienates crime victims and ignores the feelings and input of the local community. The specialization and bureaucratization of investigation have the net effect of reducing the involvement of the community and its people, thus reducing the effectiveness of the crime prevention and detection effort. The obsession of the police with fighting individual crimes, further undermines the police-community cooperation in developing a comprehensive and integrated crime control strategy (Zhang, 1997:13).

SUMMARY

This chapter investigated the philosophy, law, and practice of rural community policing in the PRC in order to provide a basis for comparative discourse in worldwide crime prevention trends and methods. It shows that, even though community policing has recently been introduced as an innovative crime control technology in urban centers in the west, community social control was practiced as a matter of routine in preindustrialized societies, as in the case of Imperial China.

The law and practice of community policing in rural China exhibits much continuity with the past. This chapter shows that community policing, as presently practiced in rural China, has deep historical roots, a venerable philosophical tradition, and a vibrant political-ideological base. In China, social control has always been local, social, and informal. The state, law, and administration were rarely involved and were used only as a last resort. Confucius thought it wise to allow the family and clan to provide education and discipline for citizens. The Imperial administrators considered it expedient to allow the local communities to police their own members. Communist ideology called for the "mass" to participate in self-government as a matter of first principle.

However, traditional community-oriented, family-based social control in China is fast becoming an endangered species. Increasingly, it is being replaced by "modern," "scientific," and "professional" policing directed from afar and imposed from without. Economic reform has gradually destroyed the fabric of the traditional rural community and its attending social control system, replacing it with a bureaucratized and impersonal police force that tries, ever so unsuccessfully, to hold on to a historical tradition and political ideology of indigenous communal rule with the reorientation of the SDCs and the introduction of community policing.

Community policing is here to stay. People's policing is gone forever!

ENDNOTE

1. This chapter is based on data collected for an ongoing research project studying law and social control in China under the auspices of the Chinese Law Program at the Chinese University of Hong Kong.

CHAPTER 16

WORLD PERSPECTIVE ON CRIME PREVENTION: A COMMUNITY POLICING APPROACH

Peter C. Kratcoski
Dilip K. Das
Arvind Verma

> Community policing is both a philosophy and an organizational strategy that allows the police and community residents to work together in new ways to solve the problems of crime (Trojanowicz and Bucqueroux, 1990:xiii).

The beginnings of modern policing emerged from an effort to bring the police and the community together for the primary purpose of preventing crime. In the early 1800s, Robert Peel's reorganization of the London Metropolitan Police was based on the mission of crime prevention (Walker, 1998:21). Community policing, however, is a new paradigm. It is an orientation that provides a complete cohesive organizational plan for modifying police work to achieve effective crime prevention (Oliver and Bartgis, 1998).

This relationship between community policing and crime prevention was explored during the Fifth International Police Executive Symposium, held at The Hague, The Netherlands, from June 2 through June 5, 1998. Hosted by Jan Wiarda, Chief of the Regional Police Force, Haaglanden, the theme of the symposium was *Crime Prevention: A Community Policing Approach.* It was attended by representatives from 21 countries. Topics specifically addressed by the participants included

1. The concept, philosophy, and history of crime prevention and community policing
2. Existing crime prevention and community policing projects, and police-community partnership initiatives

3. The mechanisms, challenges, problems, and difficulties encountered implementing crime prevention and community policings
4. Plans for future crime prevention and community policing programs

This chapter summarizes the material provided in the oral presentations and the chapters in this symposium publication. (Undated author citations in this chapter refer to symposium presentations or to chapters from the symposium book.)

THE PHILOSOPHY OF CRIME PREVENTION AND COMMUNITY POLICING

The notion of community policing and its relationship with crime prevention has been a matter of much debate among scholars (Cordner, 1995; Friedmann, 1992; Oliver and Bartgis, 1998; Trojanowicz and Bucqueroux, 1990). Leighton (1994:i) stated that "prevention of crime and the solution of crime problems could only be accomplished by a partnership between the police and the community." Carter and Radelet (1998:54) view community policing as

> ... a philosophy, not a tactic. It is a proactive, decentralized approach to policing, designed to reduce crime, disorder, and fear of crime while also responding to explicit needs and demands of the community.

Undoubtedly, "the public expects the police to prevent crime" (Moore, 1992:6), but the forms of crime that are to be prevented, the strategies to control them, and the manner in which the police ought to be organized to achieve these objectives remain contested. Clearly, crime prevention ought not to be identified only with police measures. It also must involve efforts to improve citizens' quality of life. Leighton (1994:ii–iii) argues that "improving social conditions also reduces and prevents crime by ameliorating those underlying causes which foster motivations for engaging in criminal activities." At the symposium, the key concepts of "crime prevention," "community," and "community policing" were not always perceived in exactly the same way.

Crime Prevention

Clarke's (1983) "situational crime prevention" encompasses the various conceptions of crime prevention presented. *Situational prevention* comprises actions

> (1) Directed at highly specific forms of crime, (2) that involve the management, design, or manipulation of the immediate environment ... (3) so as to

> reduce the opportunities for crime and increase the risks as perceived by a wide variety of offenders (Clarke, 1983:225).

In essence, situational prevention entails any action that attacks a specific form or instance of crime or incivility.

Johnson and associates (1993:3) point out that it is very difficult to engender crime prevention or community action in disorganized areas. They emphasize that "social crime prevention aims at mobilizing communities or young people themselves to combat the likelihood of people attempting to commit crime." This point was emphasized by several symposium participants who asserted that crime prevention must begin with educating young people that deviant behavior is wrong. Many traditional forms of social control, such as the family, religion, educational institutions, and the community have declined in importance. Other informal means of social control, such as fear of being disliked or avoided or of being shamed or punished, have also declined.

Several speakers emphasized that, if the community is to have a major impact on crime prevention, the citizens must have a value system grounded in the beliefs that it is wrong to violate the laws, that the rights of others must be respected, and that those who do not accept these premises and do violate the laws should be punished. Wiarda pointed out that the very nature of modern society, with the continuous growth of urbanization and influences of mass media, transportation, and communication, will lead to a further weakening of family, religious, school, and community controls. This loss of control will necessitate greater involvement of official and semiofficial public agencies, such as the police and social services. The methods used by governments to control and prevent crime can be very repressive, or they can enlist the aid and cooperation of the citizens.

The symposium speakers emphasized that, despite differences in customs, traditions, and values, crime prevention efforts in the various countries had many common features. It also was brought out that certain nations have borrowed or adapted effective crime prevention programs from other countries.

Community Policing

The term "community policing" has been used in various ways. According to Carter and Radelet (1998:54), community policing appears under various names, including "Problem-Oriented Policing (POP), Community Problem Oriented Policing (CPOP), Neighborhood-Oriented Policing (NOP), [and] Target Oriented Policing (TOP)." During the symposium, there was much discussion and disagreement regarding the definitions of "community" and "community policing," and the degree to which crime prevention can be achieved through community-based policing. The definition of community is important for discussions of community policing. Trojanowicz and Moore (1988) have distinguished between geographic community and community of interest. Indeed, community of interest has important consequences for community policing, since police policy is set by special interests (Langworthy and Travis, 1994:319). However, police agencies tend to view community in terms of specific territorial boundaries, and their direct involvement in and influence on the community is limited to the area over which they have jurisdiction. This definition may become obsolete due to the advent and extensive use of devices that help form communities between people who may never have physical contact.

For most nations, the concept of community is equated with people interacting face to face. Wong observed that the concept of community policing, and policing in general, can be perceived differently, depending on the history and culture of a specific country. He quoted Bayley's (1985:7) definition of policing as "people authorized by a group of people to regulate interpersonal relations within groups through the application of physical force." In a sense, this means that the concept and philosophy of community policing put people in control of their own neighborhood. Wong noted that community policing as a philosophy seeks to engage the community as coproducers of law and order, with activities ranging from consulting the community to involving the community in solving its own crimes and related social problems. Community problems can be solved only with the "active involvement, support, participation, and assistance of the public" (Wong).

Social structural conditions, such as poverty, broken families, and racial and ethnic divisions give rise to crime and disorder, and the police cannot effectively change or eliminate these anomalies. Therefore, community policing involves both a philosophy and specific approaches to policing. As a philosophy, it is based on the assumption that the public can have considerable input in preventing and controlling crime and in improving the quality of life. Police practices, grounded in community policing, must include the citizens in their planning and execution.

Approaches to community policing in the countries represented at the symposium varied considerably, depending to a great extent on the cultures, traditions, and assumptions about the nature of human beings held in the various countries. In Hong Kong (Wong), community policing and crime prevention rest on the assumptions that people are basically good and that those who deviate can be changed and brought back into the fold. The government, and specifically the police, is there to serve the people, but the people have a responsibility to assist the government and the police. Thus, the people are serving themselves.

Community policing includes a wide range of practices, such as foot patrol, bicycle patrol, storefront or ministations, neighborhood watch, and the use of volunteers. Several symposium participants mentioned that their community policing was modeled after the SARA problem-oriented policing approach. This model includes *S*canning (identifying an issue and determining whether it is a problem), *A*nalysis (collecting information on the problem from all available sources), *R*esponse (using the information to develop and implement solutions to the problem), and *A*ssessment (determining whether the response to the problem was effective) (Eck and Spelman, 1987). Hickman noted that SARA does not necessarily include community participation in crime prevention and problem-solving activities. The community policing model developed and utilized in Edmonton, Canada, however, is structured to include considerable citizen participation (Hickman). The Neighborhood Foot Patrol Program in Edmonton is designed with constables patrolling small geographic areas. Each patrol area has a storefront office used to promote community involvement and provide a place for volunteers to work. Community liaison committees include neighborhood leaders and focus on community problem solving. Both short- and long-range problem-solving strategies are employed by the constables (Hornick et al., 1989).

Similar problem-solving approaches are utilized in Israel and The Netherlands. Israeli programs identify a particular public nuisance and focus public awareness on it (Geva). Members of the public, police, and other experts analyze the nuisance and devise methods for tackling the problem that combine the resources of all relevant agencies. After implementing the plan, feedback from the public and predetermined measures of success are gathered in order to make necessary modifications before further implementation is undertaken (Geva).

In The Netherlands, community policing is grounded in the needs of the community (Wiarda). The universal concepts of community policing are adapted to local circumstances since the legitimacy of police power rests with the people. Community policing involves action learning, action research and partnerships established with citizens, local governments, and public and private organizations. The emphasis is on problem solving, with the problems defined from the citizens' perspective. A balance between obedience to the law and service to the people is a cornerstone of community policing (Wiarda). The particulars of the programs are not as important as the fact that crime prevention efforts are part of an integrated model (Mlicki).

EXISTING CRIME PREVENTION PROJECTS

Some countries have had traditions of policing that, by necessity, included citizen involvement and cooperation. Others have developed and implemented comprehensive plans. A discussion of these traditions and plans follows.

In Norway, a homogeneous, predominantly rural country of fewer than 5-million inhabitants, the police are organized around a civilian model. The police are decentralized, and all officers are trained to be generalists. They interact with the public as a normal routine, reflecting the democratic and humanistic principles of the society. They are part of the community and subject to the criticism of the local community. Crime prevention activities include both government agencies and the citizens. For example, the Norwegian Ministry of Justice and Ministry of Education collaborate with citizens to work on crime prevention activities in the schools. These programs are oriented toward both crime prevention and education on cultural diversity and respecting the rights of minorities. The police have established good working relations with emergency hospitals, health agencies, welfare and children's services agencies, and other local groups concerned with improving the quality of life of the citizens. The country has the traditions, customs, value system, and political structure to be effective in its mission (Gjefsen).

The Netherlands has 25 regional police forces under the jurisdiction of the Minister of Justice (for criminal offenses) and the Minister of the Interior (for maintaining public order). The police have a great deal of autonomy in the development of policies and procedures pertaining to the prevention and control of crime. At the local level, the mayor of the largest municipality in each police region oversees the development and implementation of policies pertaining to public order. Thus, the mayor is, in essence, the police manager, and the police chief is subordinate to the mayor. The thrust toward community policing came about after a period of rapid increases in crime. The *wijk* (neighborhood station) was the first visible community policing project. The comprehensive crime prevention neighborhood approach, is the result of gradual development, planning, implementation, and research on effectiveness. Primary, secondary, and tertiary crime prevention, coordinated by both the Ministry of the Interior and the Ministry of Justice, include neighborhood police stations, the neighborhood crime watch, neighborhood prosecution departments, enforcement teams directed at specific crime problems (such as drugs or prostitution), juvenile delinquency reduction programs, client tracking systems, and job training programs for youths. These efforts are all grounded in the concept of citizen involvement and cooperation in their planning and implementation (Aronowitz; Mlicki; Wiarda).

The Israeli police are highly centralized and authoritarian, but because of increases in crime in general, the influx of numerous immigrant groups, and evidence of a rise in serious crimes (such as racketeering and drug trafficking), a major crime prevention–community policing endeavor was launched in 1995. Full-time community police officers located in small communities maintain close contact with the local citizens. The officers use the local media to help educate the public on crime prevention strategies, involve young people in quality-of-life matters, assist them in finding solutions to problems, and mobilize neighborhood resources. Internal police efforts involve training officers in the use of mediation for handling neighborhood disputes and developing crime prevention strategies through environmental design. The police provide aid to the elderly and disabled in conjunction with volunteers. Similarly, in conjunction with the Ministry of Education, the police give advice and instruction to students on ways of preventing crime. Other programs address domestic violence, auto theft, and teenage alcohol consumption (Geva). Community policing centers are the major coordinating points for the programs; they employ the problem-solving approach described earlier.

Murray notes that the situation in Ireland is ideally suited for establishing crime prevention programs based on the community policing concept. The country has one centralized force, with all officers trained at the same academy. The Irish police have traditionally operated under the concept of "local knowledge," that is, know the area, the people, and the criminal element. Citizens and police cooperate on matters such as reducing opportunities for crime through use of closed-circuit television, neighborhood watch, local radio stations, and citizen education regarding the welfare of the community. The approach to obtaining citizens' involvement varies. For example, neighborhood watch might work best in urban areas, while community radio stations that broadcast messages might be more useful in rural areas. School programs oriented toward education in personal safety would be structured differently from those for colleges and universities. Community police officers in rural areas get to know the community, visit households and schools, and attend meetings. The officer assesses the policing needs of the community and works on specific problems (Murray).

Canada also has a long tradition of crime prevention through community policing (Lindsay). Formal, organized policing in Canada is less than 150 years old. The Northwest Mounted Police relied on the support of citizens who were interested in maintaining lawful communities (Lindsay). Their policing was grounded in a dependence on citizens for assistance, and the inhabitants were willing to help because of the mutual trust that had been established. The adoption of the professional model of policing, urbanization, growth of the welfare state, and the high mobility of the citizenry have contributed to a decline in trust and the cooperative spirit between the police and the citizens. Today, most Canadian crime prevention operates under private auspices. Programs conducted by the public sector are designed to protect children, protect property, and protect self (Lindsay). The role of the police in crime prevention is to work with citizens on projects that will lead to an improvement in the quality of life for everyone. Community policing focuses predominantly on crime prevention, and prevention is integrated into general front-line police work. Canadians believe that community policing programs must be grounded in the culture and traditions of the people and based on joint cooperative efforts of the police and citizens.

Austria, with a centralized national police force, has a tradition of involving citizens in crime prevention. Crime prevention programs and activities include lectures, individual counseling, observation and analysis of crime locations, preparation of documentation on different topics, comprehensive services (such as personal protection, object protection, and behavior-oriented advice), research and analysis of crime causes, cooperation with other institutions (such as schools and municipalities), participation in commissions and working groups, self-defense courses, security technology, and attention to problem groups (such as youth gangs and "soccer hooligans") (Edelbacher). The community policing concept is new to Austria and has been introduced on a limited basis. The Vienna Safe City Project allows community policing units to develop their own individual approaches, as long as they are oriented toward preventing crime through cooperation with the community. The five principles of community police work include joint security, lack of officer anonymity (officers use their own names), linkups with other institutions, decentralization, and constant contact with the population (Edelbacher).

In the United Kingdom, policing has traditionally been based on the concept that policing is a local concern. The police are there to serve the citizens of the community, and the citizens have traditionally been involved in crime prevention processes. A nationwide British crime prevention system includes such features as neighborhood crime watch, use of advanced technology, coordination of crime prevention programs, government research on the effectiveness of such programs, and many other efforts (Bond).

In Singapore, crime prevention is based on the assumptions that a strong police department is needed and that citizens have a responsibility for both property and personal crime prevention (Chin). Currently, neighborhood watch zones underscore community cooperation and encourage communities to solve their own problems. Many private organizations are involved in crime prevention. The strategies applied differ, depending on the group and specific interests being addressed. Many of these programs are funded by the National Crime Prevention Council. The police constable's basic training emphasizes that crime prevention is the primary duty of the police, with detection and apprehension of criminals being secondary (Leong).

In Hong Kong, the crime prevention and control model is based on the assumptions that people are basically good and desire social order and that although the government has the responsibility for taking care of the people, citizens must also contribute toward their own protection. The primary control unit is the citizen, and much crime prevention is at the grassroots level with the government serving as the last resort for control. Because of rapid changes in technology and communications, and the mobility of the population, more formal crime prevention models are being introduced, with the police and the government assuming more responsibility (Wong).

For another group of countries, the establishment of crime prevention strategies based on community policing is more difficult for a variety of reasons. In the United States there is no national police force, and there are literally thousands of small, independent departments. The concept of a highly professional, effective police department has excluded citizen involvement in most aspects of police activity. In urban areas with many minority groups, the level of communication and cooperation between the police and the citizens is almost nil. In general, community policing is implemented haphazardly, and many departments are still committed to a traditional policing model. In U.S. cities, one can find a variety of crime prevention programs, including neighborhood watch, school liaison, programs to contact and assist the elderly, victim assistance programs, and many others. The federal COPS (Community-Oriented Policing) program provides local communities with opportunities to experiment with various crime prevention strategies (Lab). Several symposium speakers pointed out that programs in their countries are patterned after those used in U.S. cities.

Numerous countries in Asia and Africa emerged from their former colonial status within the past 50 years (such as India, Kenya, Sri Lanka, and Nigeria). Many of these nations have complex and heterogeneous populations. For example, in India there are at least 18 distinct languages with more than 5,000 dialects. In addition, all the major world religions have large representations. Within each country, customs and traditions vary, and there are considerable differences in the lifestyles of those living in the urban areas compared to those in rural areas. Crime prevention and control in the precolonial period were handled within the family or by the village elders. During the colonial period in India, the local elite and the landholders were made responsible for controlling crime (Yang, 1985). In Kenya, people defended themselves, and the existence of many customs and taboos led most of the inhabitants to adhere to the norms of society voluntarily—out of fear. In Sri Lanka, offenses were considered to be sins, and religion was a strong controlling force (Goonetilleke). With the advent of colonialism, policing became more professionalized, and the police were predominantly concerned with protecting the property and interests of the landholders and businesses (Ebbe, Goonetilleke, Mwangangi, Raghavan, and Sankar). Today, many of the traditional forms of crime prevention remain intact.

Postcolonial Kenya is still very traditional: Customs, taboos, and traditions are used to induce people to conform to community standards.

Crime prevention efforts concentrate on the causes of crime, such as poverty, unemployment, and illiteracy. Much crime—particularly vehicle theft, illegal firearm distribution, and drug trafficking—is related to the economic conditions in the country. The police devote most of their personnel and resources to combating crimes that are perceived as threatening the well-being of the entire country. Some communities have developed their own neighborhood watches and established their own police stations (Mwangangi).

At the present, crime prevention in Nigeria, particularly in the rural areas, is very similar to that which existed before the colonial period. The head of the household assumes responsibility for protecting his home, property, and family. This is done by being present to protect one's property and by establishing physical barriers. The formal policing that began with the colonialization of the country by the British was predominantly concerned with protecting the economic interests of the colonial administrators and foreign landowners. During and since the colonial period, the traditional forms of crime prevention still prevailed in the rural areas. The Nigerian police engage in crime prevention tactics through patrol of the highways and the search for stolen vehicles, drug trafficking, and instances of other high-profile crimes. One important aspect of crime prevention involves progressive youth organizations that prohibit criminal behavior by their members. A person with a criminal record is not allowed to join such an organization, and if a member is convicted of a crime, he will be excommunicated from the group (Ebbe).

In India, since many people consider the police to be inadequate and corrupt, those who can afford to pay for private security will do so (Nalla, 1998). Private security guards, closed-circuit television and neighborhood watch programs are common. Police prevention activities include community patrol, the use of surveillance, and apprehension and short-term detention of high-risk individuals. Former criminals must report their movements within the country, and emigrants from other areas must have identification cards. The defining characteristic of the Indian police is order maintenance rather than crime control (Raghavan and Sankar).

Sri Lanka, a developing country where most of the population is economically disadvantaged, has experienced increases in violent crime, particularly in urban areas. A sense of community and the willingness to get involved with crime prevention is not easily aroused. Cooperation with the police and the sharing of crime information normally occurs in the rural areas, but it is more difficult to obtain in urban areas. Formal crime prevention efforts of the police consist of developing publicity campaigns to encourage citizen cooperation with the police (Goonetilleke).

Several other nations — Hungary, Russia, Romania, and Serbia — can be characterized as emerging democracies. Each had policing systems that were repressive, harsh, politically controlled, and feared by the citizens. The police were predominantly concerned with whatever was a threat to the welfare of the state (political crimes) and relied on their own investigative efforts rather than seek public cooperation. Citizens developed an orientation of noninvolvement with and fear of the police. Thus, in contrast to countries where police-citizen cooperation has a long-standing tradition, the new democracies of Eastern Europe must deal with an ingrained suspicion of and fear of the police on the part of the public.

These new democracies have approached crime prevention in various different ways. Although the police may be centrally organized, it is recognized that crime prevention strategies and programs must be geared toward the local communities. In Hungary, crime prevention efforts focus on technological developments and on police education and training. There are attempts to establish crime prevention units within the police, to change laws, and to initiate programs geared toward working with youths, different ethnic groups, and substance abusers (Banfi and Sarkozi). In Russia, crime control has historically been based on self-control and the use of repressive means to maintain order. Since the advent of democracy, one method used to deal with rising crime problems is an increased institutionalization of offenders. Rising crime in Russia appears to be directly related to the persistently depressed social and economic situation. Some crime prevention self-help networks have been established in rural areas, including agencies to assist victims of crime and sexual abuse. Neighborhood watch programs have been established and much planning for crime prevention programs is taking place (Gilinskiy).

In Romania, crime prevention activities focus on external threats to the country (the transportation of illegal goods), increases in violent and property crimes, and organized crime. The free circulation of foreigners throughout the country, along with poor social and economic conditions, are favorable to stimulating internal and external organized crime. Crime prevention efforts are directed toward controlling the international activity through networking with other nations, giving strong penalties to those convicted of drug trafficking, counterfeiting, smuggling, and trafficking in money or human beings. Rank-and-file citizens are not involved in and are not asked to assist with crime prevention activities (Mircea).

Serbia (Yugoslavia) has experienced a prolonged civil war, a severe economic crisis, an uncertain political structure, and a very serious crime problem. The majority of the citizens have no trust in the police. Current attempts at prevention involve efforts to work with youths. There are special crisis intervention programs for children, day-care centers for the young, police involvement in schools to provide lectures on drugs and the dangers of drug use, and attempts to develop crime prevention and victim assistance associations. To date, community-based policing has not been achieved (Simonovic and Radovanovic).

Central and South American countries often have centralized police systems, with the internal police being one branch of the overall military structure. Consequently, the police have traditionally served the government rather than the citizens. In Mexico, crime prevention efforts are focused on diminishing the corruption that exists within the political and police organizations. This involves prosecuting and punishing those responsible. In addition, it includes recruiting new police officers, providing them with appropriate training, and giving them incentives to make the policing profession attractive and rewarding so that they will not engage in corrupt activity. There are also attempts at more general crime prevention measures oriented toward teenagers and children, with the goal of improving their value systems and helping them take pride in being honest. There is a need to provide more opportunities for people to succeed and have a higher quality of life through legitimate means (Beller).

The peace agreement which ended the civil war in El Salvador in 1992 included elimination of the existing police and the creation of a new police force that would be professionally trained and operate with a respect for the civil rights of the citizens. A major goal of this new police force was crime prevention. Prevention of crime involved improving the efficiency of the police, working for delinquency prevention and reducing opportunities for criminal activity. Social programs that attend to the needs of youths, particularly the victims of child abuse, that care for orphans and abandoned children, and that provide programs for those who have minor substance abuse problems are prominent. The degree to which they have succeeded and the involvement of citizens in these efforts has yet to be evaluated (Orellana).

In Brazil, a major thrust of crime prevention is to attack the problem of police corruption and deal with the lack of cooperation and interaction between the police and citizens. The Latin American and Caribbean Pact for Police Protection of Human Dignity offers mechanisms for the police and the citizens to participate in developing strategies and priorities for the security and service of the public. Crucial to this goal is the provision that "police work is not limited to the accomplishment of legal mandates—but must promote human dignity" (De Camargo). The application of this theory has led to more dialogue and joint activity between the police and the communities.

Challenges to the Success of Community Policing

Although all the nations represented at the symposium have crime prevention strategies and programs in operation, not all of the strategies have been integrated into a community policing approach. Historical, cultural, or political conditions make such integration extremely difficult in some countries. Nevertheless, most of the nations are attempting some form of crime prevention that incorporates citizen involvement in prevention programs. Discussion regarding what works, what is problematic, and what is most efficient covers a number of areas. For most topics, unanimous agreement regarding appropriate solutions was not possible. The core topics included

1. A call for enlisting the assistance and cooperation of citizens
2. The need for and content of special education and training for police
3. The extent to which prevention can be replicated
4. The effects of intervention across classes of citizens
5. More extensive and meaningful research on the effectiveness of community policing

Enlisting Citizen Assistance and Cooperation

Asking for the aid, cooperation, and interest of the public in crime prevention and policing matters is a concern, even in countries such as the United States, Canada, and England where police forces were initially organized around the concept of citizen involvement (Trojanowicz and Bucqueroux, 1990:46). The reorganization of the police into a "professional" model involves a highly mechanized, bureaucratically structured, centralized department with little place for citizen involvement. In some countries, the police have always been considered the representatives of the political leaders or privileged classes. In such nations, many citizens have viewed the police with hostility, fear, and mistrust and have avoided interaction with them.

For community policing to be successful, citizens must be motivated to become involved in a meaningful way. This is true regardless of history and cultural traditions. Research has shown that police officers become ethnocentric, suspicious, and sometimes cynical, and they are reluctant to share their policing powers with citizens (Langworthy and Travis, 1994; Niederhoffer, 1969; Regoli, 1977; Skolnick, 1966). Thus, crime prevention endeavors often turn out to be public relations gimmicks designed to bolster the image of the police. If the citizens realize that the program is a sham, they will choose not to become involved.

Police are often reluctant to accept crime prevention through community policing principles, and the citizens do not want to cooperate with the police because of past negative experiences, fear, or lack of interest. Breaking down barriers of distrust and disinterest is not an easy task. In a number of countries, attempts have been made to build cooperation and communication through various public service programs. These range from providing crime prevention tips in the media, assisting with established block security and neighborhood watch programs, increasing police visibility in high-crime neighborhoods, and meeting with citizens to discuss their concerns. In many countries there is a strong effort to work with the young in the schools or the community, in order to have youths view the police in a more positive way, as well as to reduce the possibilities of youthful criminality.

Educating and Training the Police Staff

A second major concern is providing training and education for police officers so that they will become aware of the benefits of the community-based policing approach. Symposium participants did not agree on this topic. In some countries, crime prevention is the essence of police work and all officers receive the needed education and training as part of their academy work. Wiarda believes that one learns by trial and error, and thus the police should "just do it." He noted that understanding differences in culture is a very important element of community policing, since such activity must begin by discovering what is in the best interests of the citizens. Other participants believe that specific community police training should be mandatory for all officers. Aronowitz described crime prevention programs that entail police training focused on the underlying theory of community policing; that is, larger crimes can be prevented by preventing smaller ones. Developing officers' self-reliance so that they will work to solve problems is considered vital. In addition, crucial in any training is communicating the proper balance between the rights of the community and those of the individual, determining the situations that are appropriate for police involvement, and understanding diversity and the need to respect the values and traditions of racial, cultural, and religious groups.

Replicating Successful Programs

Despite the differences in customs, traditions, and values, crime prevention efforts across countries have many features in common. It is evident that there has been a considerable exchange of ideas and program models. In essence, symposium delegates asked, "Why is it necessary to reinvent the wheel?" Programs developed in the United States and England are frequently copied by other countries. Hickman noted, however, that "the learning that takes place during the design and implementation of local solutions has many more benefits than a single outcome to an individual problem.... The process is as important as the outcome." Despite the ability to exchange ideas and program designs, the implementation of crime prevention or community policing must be adapted to the specific social and cultural traditions and circumstances of the target location. Similarly, although a general training or educational approach can be developed, specific training and policies must take into account local conditions, laws, and politics.

Another major concern with adopting programs developed in other countries is the fact that, more often than not, the success of the programs has not been rigorously evaluated. Various participants emphasized the need to thoroughly understand the programs that are being borrowed, the way in which they operate, and their strengths and weaknesses. Traveling to other countries to study and observe crime prevention programs before determining whether the programs are feasible for replication is an important step.

Providing Information about Crime Prevention Programs

In many countries, particularly the more developed ones, the police use the mass media to provide the public with information and training in crime prevention. Private industries and organizations whose profits are derived from selling the services that will prevent or reduce crime have flourished. These agencies use the mass media to make the public feel that there is a need for their services or equipment. Lindsay notes that a latent effect of the media is the creation of a fear of crime that exceeds the actual crime situation. People may change their behavior (such as staying away from certain locales or at certain times), or they may take precautions such as purchasing locks, lights, firearms, and alarm systems or taking self-defense training.

The public's distorted perception of the dangers ordinary citizens face also leads to demands for more aggressive crime control tactics, higher crime clearance rates, and the targeting of specific groups as sources of the crime problem. Lindsay points out that private crime prevention programs are often controlled by the dominant and advantaged classes at the expense of the less privileged. The challenge to policing and crime prevention is to include the "have-nots" in crime prevention strategies and avoid the unintended consequences of maintaining the status quo.

Conducting Valid Research

The amount of good evaluative research on crime prevention and community policing activities is quite limited. Isolated studies on individual projects and general descriptive studies of programs are not sufficient for making policy. "The body of research on community policing has grown, but basic questions about it remain unanswered, leaving several major gaps in our current knowledge about such programs" (Lurigio and Rosenbaum, 1997).

One question deals with the extent to which the police are capable of addressing various quality-of-life goals associated with prevention activities. Issues such as employment, quality of education, recreational activities and opportunities, the availability of adequate housing, and good public transportation are just as important to citizens as the amount of crime.

The police have very little jurisdiction and influence over many of these factors. The police can, however, affect quality-of-life issues by interfacing with social service agencies such as housing bureaus, child welfare agencies, victim assistance agencies, and educational institutions. Eliminating or modifying high-crime locations can be addressed by enforcing housing codes, repairing street lights, collecting garbage, and taking other actions outside the realm of policing. Unfortunately, police involvement, when coupled with the efforts of other social service agencies and the public, is very difficult to measure, and requires highly complex research designs.

One also has to consider how all the crime prevention components coordinate and fit together. Gilinskiy cautioned against the overoptimism of expecting crime prevention and community policing activities to be a cure-all for the problems in many countries. For some countries, the causes of crime include extreme poverty, illiteracy, lack of opportunity, poor health, and inadequate housing. The problems may be so great that organized prevention activities and community policing efforts will likely have little effect. Only dramatic economic, social, and political changes will lead to the desired changes in the quality of life.

THE FUTURE OF CRIME PREVENTION THROUGH COMMUNITY POLICING

It must be emphasized that the countries represented at the symposium vary dramatically in their stages of development and implementation of crime prevention programs and community policing. Therefore, future activity in some countries may be equivalent to the present situation in others. In addition, many of the statements made in the chapters or in the presentations regarding the future could be more wishful thinking than actual possibility. The various future endeavors or activities offered for consideration can be grouped under the following categories:

- Improving the social and economic conditions
- Changing the role and functions of the police
- Increasing private and public cooperation

Improving Social and Economic Conditions

Crime prevention can best be accomplished by reducing or eradicating the social ills that breed crime, particularly by providing education and employment opportunities for youths. In El Salvador, there is a need to develop a national policy for the social prevention of delinquency (Issa). In Canada, many effective crime prevention programs that focus on youths are joint initiatives of social service agencies and the police (Hickman). In Russia, the prevailing economic, social, and political problems will necessitate the allocation of the limited resources for health, education, and social welfare programs. Crime prevention will be the province of private self-help activities (Gilinskiy). In Kenya, Nigeria, and other African countries, considerable efforts will be made to provide employment opportunities for the youth, with the assumption that this will lead to a reduction in crime. This approach also holds true for the newly independent Eastern European countries, as well as for nations in Central and South America.

Expanding the Role of the Police

In many countries there is mutual suspicion, distrust, and dislike between the police and citizens. Little can be accomplished unless a more positive relationship between the police and the public can be established. For example, in Yugoslavia the police must stop catering to the ruling classes and become a service organization for the people before broader citizen involvement can be expected. The outlook for accomplishing this is not bright.

In countries that have implemented crime prevention through community policing, the future consists of greater involvement and more cooperative ventures with other social service and community agencies. For example, in The Netherlands, a focus on preventing juvenile crime, creating safer neighborhoods, and preventing international organized crime will require the involvement of the local citizens and a continued emphasis on cooperative national and international ventures (Aronowitz). International cooperation is very important for the future, and it must include training of police officers, exchange of information, and joint crime prevention operations.

Increasing Cooperation between Police and Private Sector

Joint initiatives of public agencies, private crime prevention agencies, and organizations committed to specific issues will flourish and have considerable success (Hickman). For example, Mothers Against Drunk Driving (MADD) and Students Against Drunk Driving (SADD) have successfully communicated their messages and brought pressure resulting in legislation. Similarly, issue-oriented organizations have scrutinized police practices and judicial decision making, and have evaluated and publicized the records of public officials.

Participants from developing countries also expect greater involvement in crime prevention initiatives. In Kenya, efforts to control the drug problem will be coordinated through one national government agency. Similarly, fraud cases will be met through interagency and regional and international cooperative agreements. Standardized training in community policing will also be a central government initiative (Mwangangi).

Speakers from Nigeria, China, and Sri Lanka have emphasized that crime prevention will continue to be an important concern for individuals, families, and the community. Establishing strong moral values and respect for the law, people, and the property of others are urgent matters for each individual and family. Ebbe noted that in Nigeria and other African countries, people avoid criminal activity because they do not want to disgrace themselves, their families, and their village. However, in urban areas, such informal social control mechanisms are not as strong and there is an increasing reliance on private security measures.

SUMMARY

"The history of modern policing is littered with the remains of promising ideas that faltered and died" (Trojanowicz and Bucqueroux, 1990:357). Is community policing headed down the same path? Clearly, the concept is spreading across the globe. Police agencies, whether in developed or developing nations, are attempting to reach out to the people. The failure of the police to truly incorporate community policing concepts or to develop well-conceived crime prevention strategies fuels the trend toward private protection. The tragic consequence is that those who cannot afford to pay for security are left unprotected.

The crux of the problem is the inability of police organizations to change themselves. Police organizations and personnel are resistant to change. For many officers, community policing is "social work" done to seek media publicity. Furthermore, for departments that have a history of state repression, the transformation to a community orientation is extremely difficult.

Finally, the effectiveness of crime prevention through community policing is not well established. Programs that work well in a particular community are not easily replicated, and successful strategies can be copied only after taking into account the concerns, structure, and organization of each community. Furthermore, this kind of policing requires creativity, resources, and unique skills for which police officers need special training.

APPENDIX

INTERNATIONAL POLICE EXECUTIVE SYMPOSIUM

The International Police Executive Symposium (IPES) was founded in 1994. The aims and objectives of the IPES are to bring police researchers and practitioners closer; to facilitate cross-cultural, international, and interdisciplinary exchanges for the enrichment of the profession of policing; and to encourage discussions and writing on challenging topics of contemporary importance by those engaged in police practice and research.

One of the important activities of the IPES is the organization of an annual meeting under the auspices of a police or an educational institution. To date, meetings have been hosted by the Canton Police of Geneva (Police Challenges and Strategies, 1994), the International Institute of the Sociology of Law in Onati, Spain (Challenges of Policing Democracies, 1995), Kanagawa University in Yokohama, Japan (Organized Crime, 1996), the Federal Police in Vienna (International Police Cooperation, 1997), the Dutch Police and Europol in The Hague (Crime Prevention, 1998), the Andhra Pradesh Police in Hyderabad, India (Policing of Public Order, 1999), and the Center for Public Safety, Northwestern University, Evanston, Illinois (Traffic Policing, 2000). The 2001 meeting was hosted by the Police Academy of Szcztyno, Poland, May 27–30, 2001. Its theme was Corruption:

A Threat to World Order. The next meeting is being hosted by the Police of Turkey, May 21–24, 2002, focusing on Police Education and Training.

The majority of those participating are directly involved professionally in police executive roles, but some of the participants are internationally well-known scholars and researchers in the field. These meetings have been fruitful as a way of disseminating information on all aspects of policing. During the four-day meeting, participants, researchers, and practitioners from countries located on all continents interact in structured and informal ways. They exchange views and opinions, establish contacts and friendships, and find opportunities to engage in formal and informal dialogues on matters pertaining to policing. The executive summary of each meeting is distributed to those attending and to a wide range of interested police professionals and scholars. In addition, a book of selected papers for each meeting is published. These books are in various stages of preparation with Gordon and Breach Publishers and Prentice-Hall.

Closely connected with the IPES is *Police Practice and Research: An International Journal* (PPR). The journal highlights current practices from all over the world, providing opportunities for exchanges between police practitioners and researchers; reporting the state of public safety and the resultant quality-of-life issues on global dimensions; analyzing practices that build around the world; and bridging the knowledge gap that exists regarding who the police are, what they do, and how they maintain order, administer laws, and serve their communities in different societies, including the regions beyond the frontiers of the developed nations.

The organization is directed by a board of directors representing various countries of the world (listed below). The registered business office is located at 402 East Jackson, Macomb, IL 61455, USA; Registered Agent Douglas J. March—Telephone: 309-837-2904; Fax: 309-836-2736; E-mail: mailto:mmdlaw@Macomb.com

IPES Board of Directors

Dilip Das, President: 23 Carolanne Drive, Delmar, NY 12054, USA; Telephone: 518-475-1189; Fax: 518-475-0078; E-mail: dilip.das@plattsburgh.edu

Hofrat Maximilian Edelbacher, Vice President: Rossauerlande 5, 1090 Vienna, Austria; Telephone: 43-1-313-463-6002; E-mail: Edelmaz@magnet.at

Alexander Sam Aldrich: 173 Burke Road, Saratoga Springs, NY 12866, USA; Telephone: 518-587-5026; E-mail: aaldrich72@aol.com

H. J. Dora, DGP Headquarters Hyderabad, Andhra Pradesh, India; Telephone: 91-40-354-2224; E-mail: polchiefap@hotmail.com

Horace Judson: 101 Broad Street, Plattsburgh, NY 12901, USA; Telephone: 518-564-2010; E-mail: Horace.judson@plattsburgh.edu

George Henry Millard: Praca Coronel Fernado Prestes, 115, Luz São Paulo/SP, Brazil; Telephone: 55-11-3327-7001; E-mail: cmtg@polmil.sp.gov.br

Snezana Mijovic-Das: 23 Carolanne Drive, Delmar, NY 12054, USA; Telephone: 518-475-1189; E-mail: dilipkd@aol.com

Tonita Murray: 73 Morphy Street, Carleton Place, Ontario K7C2B7, Canada; Telephone: 613-998-0883 (office); E-mail: toni.murray@cpc.gc.ca

S. M. Ngangula: P.O. Box 50103, Ridgeway, Lusaka, Zambia; Telephone: 260-125-2872; Fax: 260-125-3543

Rick Sarre: 37-44 North Terrace, GPO Box 2471, Adelaide, South Australia 5001; Telephone: 61-883-02-0889; E-mail: Rick.sarre@unisa.edu.au

Laurent Walpen: Case Postale 236, 1211 Geneva 8, Switzerland; Telephone: 41-22-427-8111; Fax: 41-22-301-3491

Dr. Jan Wiarda: P.O. Box 264, 2501 CG The Hague, The Netherlands; Telephone: 31-70-310-2002; Fax: 31-70-310-2009

Alexander Weiss: P.O. Box 1409, Evanston, IL 60204, USA; Telephone: 800-323-4011; Fax: 847-491-5270; E-mail: alweiss@kellogg.northwestern.edu

Bibliography

Abramkin, V. (1996). *In Search of a Solution: Crime, Criminal Policy and Prison Facilities in the Former Soviet Union*. Moscow: Human Rights Publishers.

Aharoni, Z. (1991). The treatment of juvenile offenders: The preventive approach. *Innovation Exchange* 2:18–19.

Albanese, J. (1990). *Myths and Realities of Crime and Justice,* 3rd ed. Niagara Falls, NY: Apocalypse Pub.

Albrecht, G., and W. Ludwig-Mayerhofer (1995). *Diversion and Informal Social Control*. Berlin: Walter de Gruyter and Co.

Aronowitz, A. A. (1997b). Progress in community policing. *European Journal on Criminal Policy and Research* 5(4):67–84.

———. (1998). Crime prevention in the Netherlands: A government's approach to crime prevention in the Netherlands' four largest cities. *Criminaliteit en Sociale Rechtvaardigheid*.

Austrian Federal Bank (1998). Public perceptions of public institutions. Vienne, Austria: Austrian Federal Bank.

Avery v. *Midland County* (1961). 390 US 474, 481.

Awuondo, C. O. (1996). Public support and the reduction of crime strategies toward enlisting public support to criminal justice agencies in order to prevent and reduce crime rates at all levels. *The Kenya Police Review Magazine* (September).

Barkan, S. (1997). *Criminology. A Sociological Understanding*. Upper Saddle River, NJ: Prentice-Hall.

Bary, W. T. D., W. T. Chan, and B. Watson (1960). *Sources of Chinese Tradition*, Vol. 1. New York: Columbia University Press.

Bayley, D. H. (1976). *Forces of Order: Police Behavior in Japan and the United States*. Berkeley, CA: University of California Press.

———. (1988). Community policing: A report from the devil's advocate. In J. R. Greene and S. D. Mastrofski (Eds.), *Community Policing: Rhetoric or Reality*. New York: Praeger.

———. (1994). *Police for the Future*. New York: Oxford University Press.

———. (n.d.). "Community policing in Australia: An appraisal." Working Paper. Adelaide, Australia: National Police Research Unit.

———, and C. D. Sheering. (1997). Future of policing. *Law and Society Review* 30:585–602.

Ben-Harush, S. (1997). Juvenile crime prevention program. *Innovation Exchange* 6:29–31.

Bennett, G. (1977). China's mass campaigns and social control. In A. Wilson, S. L. Greenblatt, and R. W. Wilson (Eds.), *Deviance and Social Control in Chinese Society*. New York: Praeger.

Beyer, L. (1991). The logic and possibilities of "holistic" community policing. In S. McKillop and J. Vernon (Eds.), *Proceedings of the Community and the Police Conference* (Australian Institute of Criminology Conference Proceedings No. 5). Canberra: Australian Institute of Criminology.

———. (1993). *Community policing: Lessons from Victoria, Australian Studies in Law, Crime and Justice*. Canberra: Australian Institute of Criminology.

Biondo, S., and D. Palmer. (1993). *Report into Mistreatment by Police 1991–2*. Melbourne: Federation of Community Legal Centres.

Birkbeck, C. (1993). Against ethnocentrism: A cross-cultural perspective on criminal justice theories and policies. *Journal of Criminal Justice Education* 4:307–323.

Black, D. (1976). *The Behaviour of Law*. New York: Academic Press.

Boerstra, E. (1997). "Drieviant" stimuleert soziale zelfredzaamheid. *Algemeen Politie Blad* 9(26 April):4–6.

Boskovic, M. (1994). *Forms of Organization and Methods of Activity of the Police in Preventing and Detecting Criminal Acts in the Policing Area*. Belgrade: GIP Kultura.

Bowers, W. J. and J. H. Hirsch. (1987). The impact of foot patrol staffing on crime and disorder in Boston: An unmet promise. *American Journal of Police* 6:17–44.

Bracey, D. (1984). Community crime prevention in the People's Republic of China. *The Key* 10:3–10.

Brady, J. P. (1982). *Justice and Politics in People's China: Legal Order or Continuing Revolution*? New York: Academic Press.

Braithwaite, J. (1989). *Crime, Shame, and Reintegration*. Cambridge: Cambridge University Press.

Brannigan, A. (1997). Self control, social control and evolutionary psychology: Towards an integrated perspective on crime. *Canadian Journal of Criminology* 39:403–431.

Briffault, R. (1990). Our localism: Part I—The structure of local government law. *Columbia Law Review* 90:1–115.

———. (1990a). Our localism: Part II—The structure of local government law. *Columbia Law Review* 90:346–456.

Brinkman, K. (1997). Utrecht kiest voor de wijkagent. *Politie Magazine* (February):8–11.

Brown, D. (1987). Fear and loathing in Darlinghurst: To neighbourhood watch or not? *Legal Service Bulletin* 12:251–2.

Brown, L. P., and M. A. Wycoff. (1987). Policing Houston: Reducing fear and improving service. *Crime and Delinquency* 33:71–89.

Bruchman, A. (1997). A model community program for school crime prevention. *Innovation Exchange* 6:32–33.

Bruinsma, G. J. N., and A. A. Aronowitz. (1997). "Samen op Weg naar Veiligheid: Een tussentijdss procesevaluatie van de beleidsuitvoering van het Grote-stedenbeleid op het terrein van de veiligheid van Amsterdam, Den Haag, Rotterdam en Utrecht, IPIT." Unpublished report. Enschede: Universiteit Twente.

Budz, D. (1998). "Beenleigh calls for service project: Police using a 'problem-solving' approach to reduce repeat calls for service." Paper presented to the ANZ Society of Criminology Annual Conference.

Bunge, M. (1983). *Treatise on Basic Philosophy*, Vol. 5: *Understanding the World*. Boston: Riedel.

———. (1995). *Sistemas Sociales y Filosofia*. Buenos Aires: Editorial Sudamericana.

Byrne, F. (1992). The community, the council and the police: A combination to reduce crime in Footscray. In S. McKillop and J. Vernon (Eds.), *Proceedings of the National Overview on Crime Prevention Conference*. Adelaide: Australian Institute of Criminology.

Canadian Centre for Justice Statistics (1994). Criminal justice processing of sexual assault cases. *Juristat Bulletin* 14.

Carter, D.L. (1995). Community policing and D.A.R.E.: A practitioner's perspective. *BJA Bulletin* (June). Washington, DC: U.S. Department of Justice.

Carter, D. L., and L. A. Radele. (1998). *The Police and the Community*, 6th ed. Upper Saddle River, NJ: Prentice-Hall.

Chan, J. (1997). *Changing Police Culture*. Cambridge: Cambridge University Press.

Chan, W. K. K. (1975). Merchant organizations in late imperial China: Patterns of change and development. *Journal of the Hong Kong Branch of the Royal Asiatic Society* 15:28–40.

Chang, C. L. (1955). The Chinese Gentry: Studies on Their Role in Nineteenth Century Chinese Society. Seattle, WA: University of Washington Press.

Chiu, S. M. (1996). "Liberty versus Civility: A Critical Review of Efficient Policing in Hong Kong." Unpublished Master's thesis, Department of Government and Public Administration, Chinese University of Hong Kong.

Chu, T. T. (1962). *Local Government under the C'hing*. Cambridge, MA: Harvard University Press.

Clarke, R. V. (1983). Situational crime prevention: Its theoretical basis and practical scope. In Tonry, M., and N. Morris (Eds.), *Crime and Justice*, Vol. 4. Chicago: University of Chicago Press.

Cohen, J. (1968). *The Criminal Process in the People's Republic of China 1949–1963*. Cambridge, MA: Harvard University Press.

Cohen, S. (1996). Human rights and crimes of the state: The culture of denial. In J. Muncie, E. McLaughin, and M. Langan (Eds.), *Criminological Perspectives*. Thousand Oaks, CA: Sage.

Collins, R. (1982). *Sociological Insight: An Introduction to Non-Obvious Sociology*. Oxford: Oxford University Press.

Conley, C., and D. MacGillis. (1996). The federal role in revitalizing commmunities and preventing and controlling crime and violence. *NIJ Journal* 231:24–30.

Cordner, G. W. (1995). Community policing: Elements and effects. *Police Forum* 5(3):1–8.

Coster, C. (1988). Neighbourhood watch in America. In D. Challinger (Ed.), "Preventing Property Crime," *Seminar Proceedings*, No. 23. Canberra: Australian Institute of Criminology.

CPN Tools. (n.d.). *http://www.cpn.org/sections/tools/models/communitarianism.html*

CPU (Community Policing Unit). (1996a). *Community Policing in Israel*. Jerusalem: Author.

———. (1996b). *Spotlight on Community Policing in Israel*, No. 1. Jerusalem: Author.

———. (1997). *Spotlight on Community Policing in Israel*, No. 2. Jerusalem: Author.

———, and the Strategic Planning Unit. (1997). *Objectives and Measures in Police Units in the Israel Police Service: Background, Theory and Practical Elements*. Jerusalem: Author.

Crawford, E. (1992). Aboriginal community and police relations throughout New South Wales. In C. Cunneen (Ed.), *Aboriginal Perspectives on Criminal Justice*. Sydney: The Institute of Criminology.

Crime and Delinquency in USSR. (1990). *Statistical Review 1989*. Moscow: Jurid. Literatura.

Crime and Delinquency 1991. (1992). *Statistical Review*. Moscow: Finances and Statistics Publ.

Crime and Delinquency. (1997). *Statistical Review*. Moscow: MVD RF, MJ RF.

Cunneen, C. (1991). Problems in the implementation of community policing. In S. McKillop and J. Vernon (Eds.), *Proceedings of the Community and the Police Conference*. Canberra: Australian Institute of Criminology.

Dai, N. (1997). A few points of reflection on the prevention and control of crime in the new era. *Theory and Practice of Public Security* 6(3):46–50.

Davis, N., and B. Anderson. (1983). *Social Control: The Production of Deviance in the Modern State*. New York: Irvington Pub.

Dixon, D. (1997). *Law in Policing: Legal Regulation and Police Practices*. Oxford: Clarendon Press.

Dixon, D. (1998). "Broken windows, zero tolerance, and the New York miracle." Paper presented to the Australian/New Zealand Society of Criminology Annual Conference.

Dölling, D., and T. Feltes. (1993). Community policing: Comparative aspects of community oriented police work. *Empirische Polizeiforschung*, Vol. 5. Amsterdam: Felix Verlag.

Donzinger, S. (1996). *The Real War on Crime: The Report of the National Criminal Justice Commission*. New York: HarperCollins.

Durkheim, E. (1933). *The Division of Labor in Society*. New York: The Free Press.

Dutton, M. R. (1992). *Policing and Punishment in China*. Cambridge: Cambridge University Press.

Ebbe, O. N. I. (1989). Crime and delinquency in metropolitan Lagos: A study of crime and delinquency area theory. *Social Forces* 67:751–765.

———. (1996). *Comparative and International Criminal Justice Systems: Policing, Judiciary, and Corrections*. Boston: Butterworths–Heinemann.

Eck, J. E., and W. Spelman. (1987) *Problem Solving: Problem-Oriented Policing in Newport News*. Washington, DC: Police Executive Research Forum.

Eck, J.E., and D.P. Rosenbaum (1994). The new police order: Effectiveness, Equity, and efficiency in community policing. In Rosenbaum, D.P. (Ed.), *The Challenge of Community Policing: Testing the Promise*. Thousand Oaks, CA: Sage Publications, Inc.

Editorial. (1979). Prospect and retrospect: China's socialist legal system. *Beijing Review* 22(2):12,25.

Esbensen, F. (1987). Foot patrols: Of what value? *American Journal of Police* 6:45–65.

Etter, B. (1995). Mastering innovation and change in police agencies. In B. Etter and M. Palmer (Eds.), *Police Leadership in Australasia*. Annandale: Federation Press.

Farjun, Y. (1997). Community policing model for the treatment of drug-abusing juvenile offenders. *Innovation Exchange* 6:27–28.

Feeley, M. M. (1997). Two models of the criminal justice: An organization perspective. In B. W. Hancock and P. M. Sharp (Eds.), *Public Policy: Crime and Criminal Justice*. Englewood Cliffs, NJ: Prentice-Hall.

Feng, S. L. (1994). *General Plan for Chinese Crime Prevention* Beijing: Falu Chubanshe.

Ferwerda, H., and A. Geutjes. (1997). Camera's in de bus: Een effectief preventiemiddel. *SEC* 11(4):10–12.

Financial News. (1992). No. 2 (Russian).

Finnane, M. (1994). *Police and Government: Histories of Police in Australia*. Melbourne: OUP.

Foran, R. W. (1962). *The Kenya Police 1887–1960*. Plymouth Eng: Clarke, Double and Brendon Ltd.

Friedmann, R. R. (1986). The neighborhood police officers and social service agencies in Israel: A working model for cooperation. *Police Studies* 9.

———. (1992). *Community Policing*. Hertfordshire, UK: Harvester Wheatsheaf.

Gal, S. (1993). The Civil Guard's adjustment to the changing needs of Israel society. *Innovation Exchange* 4:12–14.

Gallup Corporation (1993). *Results of Poll on Crime Issues*. Princeton, NJ: Author.

———. (1995). *Results of Poll on Crime Issues*. Princeton, NJ: Author.

Gao, J. (1994). Villagers' committee's organization establishment, background, current status, and policy aim. *Studies in Law* 2:11–16.

Geason, S., and P. Wilson. (1990). *Crime Prevention: Theory and Practice*. Canberra: Australian Institute of Criminology.

Geva, R. (1981). Crime prevention: Past, present and future. In Israel Police (Ed.), *Target Hardening—Preventing Property Crime*. Jerusalem: Israel Police.

———. (1990a). The founding of the National Council for Crime Reduction in Israel. *Innovation Exchange* 1:6–7.

———. (1990b). Working group on combatting crime related to vehicles. *Innovation Exchange* 1:7–8.

———. (1992). *Strategies in Crime Prevention: Techniques, Implementation and Evaluation*. Jerusalem: Ministry of Police.

———. (1995a). "The Prevention of Crime and the Treatment of Offenders in Israel: A Report." *Proceedings of the Ninth UN Congress on the Prevention of Crime and the Treatment of Offenders*, Cairo.

———. (1995c). "Effective national and international action against terrorism: The Israeli experience. In R. Geva (ed.), The Prevention of Crime and the Treatment of Offenders in Israel: A Report." *Proceedings of the Ninth UN Congress on the Prevention of Crime and the Treatment of Offenders*, Cairo.

———. (1995d). Preventing domestic violence (spouse abuse) in Israel. In R. Geva (Ed.), "The Prevention of Crime and the Treatment of Offenders in Israel: A Report." *Proceedings of the Ninth UN Congress on the Prevention of Crime and the Treatment of Offenders*, Cairo.

———. (1998). *Results of Client Satisfaction with Policing Services in Galilee Sub-District and Results of Client Satisfaction with Policing Services in Sharon Sub-District*. CPU Report. Jerusalem: Crime Prevention Unit.

———, and I. Israel. (1982). Anti-burglary campaign in Jerusalem: Pilot project update. *Police Chief* 49:44–46.

Gibbs, J. P. (1982). *Social Control*. Beverly Hills, CA: Sage.

Gimshi, D. (1995). Planning and implementing organizational changes. In Community Policing Unit (Ed.), *Community Policing in Israel: The First Steps*. Jerusalem: Community Policing Unit.

———. (1997). Moving toward community policing: Planning and implementing organizational change. *Innovation Exchange* 6:14–16.

Goldsmith, A. (1990). Taking police culture seriously: Police discretion and the limits of law. *Policing and Society* 1:91–114.

Goldstein, A. P. (1990). *Problem-Oriented Policing*. New York: McGraw-Hill.

Gönczöl, K. (1996). Deviance, control over deviance, strategies of prevention. In K. Gönczöl, L. Korinek and M. Lévai (Eds.), *Criminological Knowledge—Crime—Control of Crime*. Budapest: Corvina.

Grabosky, P. (1992). Law enforcement and the citizen: Non-overnmental participants in crime prevention and control. *Policing and Society* 2:249–271.

Graham, J., and T. Bennett. (1995). *Crime Prevention Strategies in Europe and North America*. Helsinki: HEUNI Publications.

Green, L. (1995). Policing places with drug problems: The multi-agency response team approach. In J. E. Eck and D. Weisburd (Eds.), *Crime and Place*. Monsey, NJ: Criminal Justice Press.

Greene, J. R., W. T. Bergman, and E. J. McLaughlin. (1994). Implementing community policing: Cultural and structural change in police organizations. In D. P. Rosenbaum (Ed.), *The Challenge of Community Policing: Testing the Promises*. Thousand Oaks, CA: Sage.

Gunther-Moore, L., and C. D. van der Vijver. (1997). Soziale Zelfredzaamheid van Burgers. *Algemeen Politie Blad* 146(7):14–15.

Gurwit, R. (1933). Communitarianism: You can try it at home. *Governing* 6:33–39.

Heffetz, A. (1995). Moving toward community policing. In Community Policing Unit (Ed.), *Community Policing in Israel: The First Steps*. Jerusalem: Community Policing Unit.

Henderson, R. (1972). *The King in Every Man: Evolutionary Trends in Onitsha-Ibo Society and Culture*. New Haven: Yale University Press.

Hendricks, J., and B. Byers. (1996). *Crisis Intervention in Criminal Justice/Social Service*. Springfield, IL: Charles C. Thomas.

Hickman, L. T. (1998). *Community Justice Forums: Report on Implementation*. Lethbridge, BC: RCMP.

Hinton, W. (1966). Fanshen. New York: Vintage.

Hirshi, T. (1969). *Causes of Delinquency*. Berkeley: University of California Press.

Hirschi, T. (1969). *Causes of Delinquency*. Berkeley, CA: University of California Press.

Hoefnagels, P., and I. van Erpicum (1997). *The Prevention Pioneers: History of the Hein Roethof Award 1987–1996*. The Hague: Prevention, Youth Protection and Probation Department, Ministry of Justice.

Hoffman, A. (1995). Inmate rehabilitation in Israel. In Geva, R. (Ed.), "The Prevention of Crime and the Treatment of Offenders in Israel: A Report." *Proceedings of the Ninth UN Congress on the Prevention of Crime and the Treatment of Offenders*, Cairo.

Homel, R. (1994). Can police prevent crime? In K. Bryett and C. Lewis (Eds.), *Unpeeling Tradition*. Brisbane: CAPCM.

Hoover, L. T. (1992). Police mission: An era of debate. In L. T. Hoover (Ed.), *Police Management: Issues and Perspectives*. Washington, DC: Police Executive Research Forum.

Hope, T. (1994). Problem-oriented policing and drug market locations: Three case studies. In R. V. Clarke (Ed.), *Crime Prevention Studies*, Vol. 2. Monsey, NJ: Criminal Justice Press.

Horn, J. (1993). Probleemgerichte aanpak van onveiligheid. *Justitiële Verkenningen* 19(5):24–49.

———, and E. Koolhaas. (1990). Wijkteampolitie in Nederland. *Het tijdschrijft voor de politie* 52(1):22–30.

Horner, B. (1993). *Crime Prevention in Canada: Toward a National Strategy*. Ottawa: The Standing Committee on Justice and the Solicitor General.

Hornick, J. P., B. A. Burrows, I. Tjosvold, and D. M. Phillips. (1989). *An Evaluation of the Neighborhood Foot Patrol Program of the Edmonton, Canada Police Service—*

Report Prepared for Edmonton Police Department. Edmonton: Edmonton Canada Police Department.

Horrall, S. W. (1974). The march west. In H. Dempsey (Ed.), *Men in Scarlet*. Calgary: McLelland and Stewart West.

Hovav, M. (1988). Alternatives to imprisonment. *Issues in Criminology* 3.

———. (1995). Correctional services and the prevention of crime: The Ministry of Labor and Social Affairs. In R. Geva (Ed.), "The Prevention of Crime and the Treatment of Offenders in Israel: A Report." *Proceedings of the Ninth UN Congress on the Prevention of Crime and the Treatment of Offenders*, Cairo.

Hsiao, K. C. (1960). *Rural China: Imperial Control in the Nineteenth Century*. Seattle: University of Washington Press.

Hu, H. C. (1988). *The Common Descent Group in China and Its Functions*. New York: Viking Fund.

Huang, S. M. (1986). *The Spiral Road: Change in a Chinese Village through the Eyes of a Communist Party Leader*. Boulder, CO: Westview Press.

Hungarian Parliament (1989). *Hungarian Constitution*. Budapest, Hungary: author.

Ignjatovic, D. (1995). The state of crime in the Federal Republic of Yugoslavia. *Zbornik Radova Policijske Akademije* 1:35–52.

Indian Police Commission. (1903). *Report of the Indian Police Commission*. New Delhi: author.

Inkster, N. D. (1989). *Commissioner's Directional Statement—1990*. Ottawa: Royal Canadian Mounted Police.

Inspectie Politie. (1997). *Gebiedsgebonden politiewerk, een verkenning*. The Hague: Ministerie van Binnenlandse Zaken.

Institute for Demographics (1996). *Facts and Figures, Annual Edition 1996*. Vienna, Austria: Institute for Demographics.

Intensiveren Samenwerking Raad en Politie (1998). The Hague: Jeugdcriminaliteit, Ministerie van Justitie. *http://www.minjust.nl/a_beleid/thema/jeugd/beleid/project/pr_9.htm*

Israel Prison Service. (1995). Treatment and rehabilitation of inmates. In R. Geva (Ed.), "The Prevention of Crime and the Treatment of Offenders in Israel: A Report." *Proceedings of the Ninth UN Congress on the Prevention of Crime and the Treatment of Offenders*, Cairo.

Jakovlev, A. (1971). *Crime and Social Psychology*. Moscow: Jurid. Literatura.

James, S., and K. Polk. (1989). Policing youth: Themes and directions. In D. Chappell and P. Wilson (Eds.), *Australian Policing: Contemporary Issues*. North Ryde, Australia: Butterworths.

Jasovic, Z. (1966). Prevention and reduction of delinquent behavior of the young in our country. *Jugoslovenska Revija za Kriminologiju i Krivicno Pravo* 4:559–570.

Jedelsky, P. (1998). *Kriminalpolizeiliches Vorbeugungsprogramm der Wiener Polizei*. Vienna: Author.

Jennings, J. (1974). The plains Indian and the law. In H. Dempsey (Ed.), *Men in Scarlet*. Calgary: McLelland and Stewart West.

Jeremic, B. (1963). Criminological research as a necessary precondition of crime reduction. *Jugoslovenska Revija za Kriminologiju i Krivicno Pravo* 1:3–10.

Jeugdcriminaliteit. (1998). The Hague: Ministry of Justice. *http://www.minjust.nl/a_beleid/thema/jeugd/jeugd.htm*

Jiang, Y. H. (1996). Improving upon the basic qualities of the People's police. *Gongan Yanjiu* 6:9–12.

Johnson, E. H. (1984). Neighborhood police in the PRC. *Police Studies* 6:8–12.

Johnson, V., J. Shapland, and P. Wiles. (1993). *Developing Police Crime Prevention: Management and Organizational Change*. London: Home Office Police Department.

Johnston, L. (1992). *The Rebirth of Private Policing*. London: Routledge.

Junger-Tas, J. (1997a). The future of a prevention policy toward juveniles. *European Journal on Criminal Policy and Research* 5:101–114.

———. (1997b). *Jeugd en Gezin II: Naar een effectief preventiebeleid*. The Hague: Directoraat-Generaal Preventie, Jeugd en Sancties, Ministerie van Justitie.

Justitiekrant. (1998). Actieprogramma Marokkaanse jeugd snel van start. *Ministerie van Justitie* (17 April):1.

Kadman, Y. (1995). The law for the prevention of the abuse of minors and the helpless. In R. Geva (Ed.), "The Prevention of Crime and the Treatment of Offenders in Israel: A Report." *Proceedings of the Ninth UN Congress on the Prevention of Crime and the Treatment of Offenders*, Cairo.

Karpetch, I. (1969). *Problems of Criminality*. Moscow: Jurid. Literatura.

Kelling, G. L. (1988). *Police and Communities: The Quiet Revolution. Perspectives on Policing*, no. 1. Washington, DC: National Institute of Justice.

Kennedy, D. M. (1996). Neighborhood revitalization: Lessons from Savannah and Baltimore. *NIJ Journal* 231:13–17.

Kleiman, M. A. R., and K. D. Smith (1990). State and local drug enforcement on retail heroin dealing. In M. Chaiken (Ed.), *Street Level Drug Enforcement: Examining the Issues*. Washington, DC: National Institute of Justice.

Kleiman, W. M., and G. J. Terlouw. (1997a). *Kiezen voor een Kans*. The Hague: WODC, Ministerie van Justitie.

Klockars, C. B. (1985). *The Idea of Police*. Beverly Hills, CA: Sage.

Kohnstamm, J. (1997). Crime Prevention as a local enterprise. *European Journal on Criminal Policy and Research* 5:61–64.

Kosoplechev, N., and F. Ismailova. (1997). *Crime Prevention in Regions: State, Experience*. Moscow: NII of General Prosecutor's Office RF.

Krainz, K. W. (1994). *Hauseinbrüche Schwergemacht*, 2nd ed. Vienna: University of Graz.

Krivokapic, V. (1986). Organs of internal affairs and crime prevention. *Jugoslovenska Revija za Kriminologiju i Krivicno Pravo* 1–2:109–121.

———. (1992). Basic methods and organs in crime struggling. *Bezbednost* 1:28–32.

———. (1994). Basic concepts of criminal policy thought of academician Milutinovic. *Anali Pravnog Fakulteta u Beogradu* 1–2:105–118.

———. (1996). *Criminalistics Tactics I*. Belgrade: Police Academy.

———, and M. Boskovic (1984). The role of organs of internal affairs in crime prevention and reduction. *Jugoslovenska Revija za Kriminologiju i Krivicno Pravo* 3:293–316.

Krstic, O. (1997). *Applied Criminalistics*. Belgrade: Institute for Textbooks and Educational Means.

Kudrjavcev, V. (1968). *Cause in Criminology: About the Structure of Individual Criminal Behavior*. Moscow: Jurid. Literatura.

Lab, S. P. (1984) Police productivity: The other eighty percent. *Journal of Police Science and Administration* 12:297–302.

———. (1997). *Crime Prevention: Approaches, Practices, and Evaluation*, 3rd ed. Cincinnati, OH: Anderson Publishing.

Lahav. C. (1995). Socialization of youth at risk. In R. Geva (Ed.), "The Prevention of Crime and the Treatment of Offenders in Israel: A Report." *Proceedings of the Ninth UN Congress on the Prevention of Crime and the Treatment of Offenders*, Cairo.

Langworthy, R. H., and L. F. Travis III. (1994). *Policing in America: A Balance of Forces*. New York: Macmillan.

Lee, J. F. (1988). *History of Chinese Legal System*. Taipei: Liang-jin chubanshiwu gongsi.

Leighton, B. (1994). Forward. In G. Saville (Ed.), *Crime Problems: Community Solutions—Environmental Criminology as a Developing Prevention Strategy*. Port Moody, BC: AAG Inc. Pub.

Leighton, B. N. (1994). Community policing in Canada: An overview of experience and evaluations. In D. P. Rosenbaum (Ed.), *The Challenge of Community Policing: Testing the Promises*. Thousand Oaks, CA: Sage.

Leng, S. C., and H. D. Chiu. (1985). *Criminal Justice in Post Mao China*. Albany, NY: State University of New York.

Levy, A. (1991). *From the Periphery to the Center*. Jerusalem: The Department for Children and Youth.

Li, S. J., and X. You. (1994). *Zhongguo Nongcun Jiceng Shehui Zuzhi Tixi Yanjiu*. Beijing: Zhongguo nongcun chunbanshe.

Liu, W. H. C. (1959). *The Traditional Chinese Clan Rules*. Locust Valley, NY: J.J. Augustin Incorporated, Publisher.

Liu, H. N., and Y. F. Yang. (1984). *Knowledge in Chinese Legal History*. Helungjian: Helungjian renmin chubanshe.

Loncarevic, D. (1987). Preventive function of the judicial organs. *Jugoslovenska Revija za Kriminologiju i Krivicno Pravo* 4:39–49.

Lotz, J. (1984). *The Mounties: The History of the Royal Canadian Mounted Police*. Greenwich, CT: Bison Books.

Luo, R. Q. (1994). *A Treatise on People's Police Work*. Beijing: Qunzhong chubanshe.

Lurigio, A. J., and D.P. Rosenbaum (1986). Evaluation research in community crime prevention: A critical look at the field. In Rosenbaum, D.P. (Ed.) *Community Crime Prevention: Does It Work*? Beverly Hills, CA: Sage.

Lurigio, A. J., and D. P. Rosenbaum. (1997). Community policing: Major issues and unanswered questions. In M. L. Dantzer (Ed.), *Contemporary Policing: Personnel Issues and Trends*. Boston: Butterworth-Heinemann.

Madrazo, J., and W. Beller. (1997). Los valores éticos y los derechos humanos. In S. G. Ramirez (Ed.), *Los Valores en el Derecho Mexicano*. Mexico DF: Universidad Nacional Autónoma de México.

Madsen, R. (1984). *Morality and Power in a Chinese Village*. Berkeley: University of California Press.

Mahrer, K. (1995). Community Policing in Vienna, "Zwischenbericht zum Durchbruch einer Strategie." *Die Bundesgendarmerie* (November/December).
Mao, T. T. (1971). *Selected Readings of Mao Tse Tung*. Beijing: Foreign Language Press.
Mao, Z. D. (1967). *Selected Works of Mao Zhedong*. Beijing: Foreign Language Press.
Mastrofski, S. D., and J. R. Greene. (1993). Community policing and the rule of law. In D. Weisburd and C. D. Uchida (Eds.), *Police Innovation and Control of Police*. New York: Springer-Verlag.
McDonald, J., D. Moore, T. O'Connell, and M. Thorsborne. (1995). *Real Justice Training Manual: Coordinating Family Group Conferences*. Pipersville, PA: The Piper's Press.
McGregor, D. (1960). *The Human Side of Enterprise*. New York: McGraw-Hill.
McMillan, E. (1992). An overview of crime prevention in Victoria. In S. McKillop and J. Vernon (Eds.), *Proceedings of the National Overview on Crime Prevention Conference*. Adelaide: Australian Institute of Criminology.
Miller, L. S., and K. M. Hess. (1994). *Community Policing: Theory and Practice*. Minneapolis: West Publishing.
Milutinovic, M. (1964). Crime and other phenomena of social pathology and the role of the commune in their reduction. *Pravni Zbornik* 1:9–32.
———. (1965). Prevention of adolescent delinquency. *Arhiv za Pravne i Drustvene Nauke* 1–2:61-84.
———. (1976). Social self-protection as theory, policy and practice of protection of our society and its values. *Jugoslovenska Revija za Kriminologiju i Krivicno Pravo* 1:3–26.
———. (1984). Social prevention as an area of criminal policy. *Jugoslovenska Revija za Kriminologiju i Krivicno Pravo* 3-4:393–420.
Ministerie van Justitie. (1986). *Eindrapport Commissie kleine Criminaliteit. Staatsuitgeverij's-Gravenhage*. The Hague: Author.
Ministry of Justice, Prevention (1997). *Youth Protection and Probation Department. Celebrating Crime Prevention Award 1997*. The Hague: Author.
Moore, M. H. (1992). Problem solving and community policing. In M. Tonry and N. Morris (Eds.), *Crime and Justice*. Chicago: University of Chicago Press.
Mor, R. (1990). Licensing businesses as a tool for crime prevention. *Innovation Exchange* 1:9–10.
Muga, E. (1972). Is criminality inherited?: A survey of contemporary research. *Journal of Eastern African Research and Development* 2.
Mukherjee, S., and A. Graycar. (1997). *Crime and Justice in Australia*, 2nd ed. Annandale, Australia: Hawkins Press.
Murray, P. (1998). *Commissioner's Directional Statement*. Ottawa: Royal Canadian Mounted Police.
Mushanga, T. M. (1976). *Crime and Defiance: An Introduction to Criminology*. Kampala: East African Literature Bureau.
Nalla, M. (1998). Opportunities in an emerging market. *Security Journal* 10:15–21.
National Anti-Drug Authority. (1995a). The national agency coordinating the fight against drug abuse. In R. Geva (Ed.), "The Prevention of Crime and the Treatment of Offenders in Israel: A Report." *Proceedings of the Ninth UN Congress on the Prevention of Crime and the Treatment of Offenders*, Cairo.

———. (1995b). The mass media and drug abuse prevention. In R. Geva (Ed.), "The Prevention of Crime and the Treatment of Offenders in Israel: A Report." *Proceedings of the Ninth UN Congress on the Prevention of Crime and the Treatment of Offenders*, Cairo.

National Crime Prevention Council. (1997). Compendium of Approaches. *http://www.crime-prevention.org/ncpc/ publications*

———. *Mobilizing for Action: The Second Report*. Ottawa: NCPC.

National Crime Records Bureau. (1997). *Crime in India 1995*. New Delhi: Author.

National Police Commission. (1977). *Report of the National Police Commission*. New Delhi: Author.

———. (1980). *Report of the National Police Commission*. New Delhi: Author.

———. (1981). *Report of the National Police Commission*. New Delhi: Author.

Neal, A. G. (1971). Alienation and social control. In J. P. Scott and S. P. Scott (Eds.), *Social Control and Social Change*. Chicago: University of Chicago Press.

Niederhoffer, A. (1969). *Behind the Shield*. Garden City, NY: Doubleday.

Nir, T. (1989). *The Beit-Dagan Community Program to Fight Drugs*. Jerusalem: Ministry of Police.

Nixon, C. (1992). New South Wales: Crime prevention directions for the future. In S. McKillop and J. Vernon (Eds.), *Proceedings of the National Overview on Crime Prevention Conference*. Adelaide: Australian Institute of Criminology.

O'Connor, M. (1988). Community policing in New South Wales. *Legal Service Bulletin* 13(2):52–55.

Offens, H. (1997). Meer duidelijkheid rond cameratoezicht. *SEC* 11:26–27.

Okola, O. J. K. (1996). "Patterns, trends and extent of person to person violence and property crime in Mombasa town: A victimization study, 1985–1981." Unpublished master's thesis, Department of Sociology, University of Nairobi.

Okonkwo, C. O. (1966). *The Police and the Public in Nigeria*. London: Sweet and Maxwell.

Oliver, W. M., and E. Bartgis. (1998). Community policing: A conceptual framework. *Police Studies* 21:490–509.

O'Malley, P. (1979). Class conflict, land and social banditry: Bushranging in nineteenth century Australia. *Social Problems* 26:271.

O'Malley, P. (1994). Neo-liberal crime control: Political agendas and the future of crime prevention in Australia. In D. Chappell and P. Wilson (Eds.), *The Australian Criminal Justice System: The Mid 1990s*. Sydney: Butterworths.

Parish, W. L., and M. K. Whyte. (1978). *Village and Family in Contemporary China*. Chicago: University of Chicago Press.

Patterson, D. (1988). Police in the P.R.C. *Police* (October):26–29.

Peak, K. J., and R. W. Glensor. (1996). *Community Policing and Problem Solving*. Englewood Cliffs, NJ: Prentice-Hall.

Peng, Z. (1991). *A Discussion on Political-Legal Work in New China*. Beijing: Zhongyan wenxian chubanshe.

Pepinski, H., and R. Quinney. (1991). *Criminology as Peacemaking*. Bloomington: Indiana University Press.

Pepinsky, H. E. (1973). The people v. the principle of legality in the People's Republic of China. *Journal of Criminal Justice* 4:51–60.

———. (1975). Reliance on formal written law, and freedom and social control in the United States and the People's Republic of China. *British Journal of Sociology* 26:330–42.

Philip, P. (1998). Community policing: An international blueprint. *International Police Review* (March\April).

Pihler, S. (1997). Dilemmas regarding an external control of the police. *Glasnik Advokatske Komore Vojvodine* 11:338–392.

Police Foundation. (1981). *The Neward Foot Patrol Experiment.* Washington, DC: Author.

Police Laws, Zhongguo Jingcha Falu Fagui Shiyi Daquan Editorial Committee. (1993). *A Compendium of Chinese Police Laws, Regulations, and Interpretations.* Beijing: Jingguan Jiaoyu chubanshe.

Police Laws, Zhonghua Renmin Gongheguo Falu Lifa Sifa Jieshi Anli Daquan Editorial Committee. (1990). *Compendium of PRC Laws, Regulations, Judicial Interpretations, and Cases.* Hebei, China: Heibei chubanshe.

Prenzler, T., and R. Sarre. (1998). Regulating private security in Australia. In *Trends and Issues in Crime and Criminal Justice.* Canberra: Australian Institute of Criminology.

Quah, S., and J. S. T. Quah. (1987). *Friends in Blue: The Police and the Public in Singapore.* Singapore: Oxford University Press.

Raghavan, R. K. (Forthcoming) *Policing a Democracy: The Case of India and the U.S.* New Delhi: Manohar.

Regev, Y., and R. Shoham. (1998). "Comparison of police-officers' attitudes in a 'community policing station' and a noncommunity policing station." Unpublished research report.

Regoli, R. M. (1977). *Police in America.* Washington, DC: University Press of America.

Reid Report. (1990). *The Reid Report* 5 (May).

Reiner, R. (1992). *The Politics of the Police,* 2nd ed. Toronto: University of Toronto Press.

Reiss, A. J. (1983). Crime control and the quality of life. *American Behavioral Scientist* 27:43–58.

Roberg, R. R. (1994). Can today's police organizations effectively implement community policing? In D. P. Rosenbaum (Ed.), *The Challenge of Community Policing: Testing the Promises.* Thousand Oaks, CA: Sage.

Robinson, J. (1989). Police effectiveness: Old dilemmas, new directions. In D. Chappell and P. Wilson (Eds.), *Australian Policing: Contemporary Issues.* North Ryde, Australia: Butterworths.

Rosen, M. S. (1990). Law enforcement news interview: Lawrence Sherman. *Law Enforcement News* (March 31):9–12.

Rosenbaum, D. P., et al. (1991). Crime prevention, fear reduction, and the community. In W. A. Geller (Ed.), *Local Government Police Management,* 3rd ed. Washington, DC: International City Management Association.

Rothwax, H. (1996). *Guilty: The Collapse of Criminal Justice.* New York: Random House.

Royal Canadian Mounted Police. (1992). *Futuristic Management Concepts in Community Policing: Action Plan.* Lethbridge, Alberta: RCMP.

———, Administration Manual (1994). *Principles of Policing and Management in the R.C.M.P.* (A.M. 94-11-25). Ottawa: RCMP.

———, Corporate Management, Audit, Evaluation and Corporate Services Directorate (1995a). *Community Policing Review.* Ottawa: RCMP.

———, Pony Express (1995b). *The Cadet Training Program.* Ottawa: RCMP.

———, Public Affairs and Information Directorate (1996). *R.C.M.P. Fact Sheets 1996.* Ottawa: Minister of Supply and Services Canada.

———, Community, Contract & Aboriginal Policing Services Directorate (1998). *Regional Community Policing Performance Review–1998.* Ottawa: RCMP.

Royal Commission. (1991). *Royal Commission into Aboriginal Deaths in Custody, National Report.* Canberra: AGPS.

Russian Statistic Annual (1995). *Statistic Review.* Moscow: Goskomstat.

Sadd, S., and R. Grinc. (1994). Innovative neighborhood oriented policing: An evaluation of community policing in eight cities. In D. P. Rosenbaum (Ed.), *The Challenge of Community Policing: Testing the Promises.* Thousand Oaks, CA: Sage.

———, and R. Grinc. (1996). Implementation challenges in community policing: Innovative neighborhood-oriented policing in eight cities. *NIJ Research in Brief.* Washington, DC: U.S. Department of Justice.

Sampson, A., P. Stubbs, D. Smith, G. Pearson and H. Blagg. (1988). Crime, localities and the multi-agency approach. *British Journal of Criminology* 28:478–493.

Sarre, R. (1991a). Political pragmatism versus informed policy: Issues in the design, implementation and evaluation of anti-violence research and programs. In D. Chappell, P. Grabosky, and H. Strang (Eds.), *Australian Violence: Contemporary Perspectives.* Canberra: Australian Institute of Criminology.

———. (1991b). Community policing: Success or failure? Exploring different models of evaluation. In S. McKillop and J. Vernon (Eds.), *Proceedings of the Community and the Police Conference.* Canberra: Australian Institute of Criminology.

———. (1992). Problems and pitfalls in crime prevention. In S. McKillop and J. Vernon (Eds.), *Proceedings of the National Overview on Crime Prevention Conference.* Adelaide: Australian Institute of Criminology.

———. (1994). The evaluation of criminal justice initiatives: Some observations on models. *The Journal of Law and Information Science* 5:35-46.

———. (1996). The state of community-based policing in Australia: Some emerging themes. In D. Chappell and P. Wilson (Eds.), *Australian Policing: Contemporary Issues,* 2nd ed. Sydney: Butterworths.

———. (1997). Crime prevention and police. In P. O'Malley and A. Sutton (Eds.), *Crime Prevention in Australia: Issues in Policy and Research.* Annandale, Australia: Federation Press.

———. (1998). Accountability and the private sector: Putting accountability of private security under the spotlight. *Security Journal* 9:97–102.

Shalev, O., and P. Yehezkeli. (1995). "The Be'er Sheva multi-organizational treatment model for spouse abuse. In R. Geva (Ed.), "The Prevention of Crime and the Treatment of Offenders in Israel: A Report." *Proceedings of the Ninth UN Congress on the Prevention of Crime and the Treatment of Offenders*, Cairo.

Shapira, N. (1990). Summary of the activities of the working group on combatting crimes against the elderly. *Innovation Exchange* 1:8–9.

Sheng, Y. Z. (1997). A discussion of police work methods of community police officers. *Gongan Yanjiu 1997*(4):29–31.

Sherman, L. (1974). The sociology and the social reform of the American police: 1950–1973. *Journal of Police Science and Administration* 2(3).

———. (1990). Police crackdowns: Initial and residual deterrence. In M. Tonry and N. Morris (Eds.), *Crime and Justice*, Vol. 12. Chicago: University of Chicago Press.

———. (1992). Book review of Herman Goldstein's "Problem-Oriented Policing." *Journal of Criminal Law and Criminology* 82:690–707.

Shimshi, E. (1997). Crime prevention: Who should care? *Metzila* 1.

Skakavac, Z. (1997). Some aspects of the passing and implementation of the instructions on the manner of organization and implementation of internal affairs in the area of security. *Bezbednost* 5:633–639.

Skogan, W. G. (1990). *Disorder and Decline: Crime and the Spiral Decay in American Neighborhoods*. Berkley: University of California Press.

———. (1995). Community policing in Chicago: Year two. *NIJ Research Preview*. Washington, DC: U.S. Department of Justice.

———, and M. G. Maxfield (1983). *Coping with Crime*. Beverly Hills, CA: Sage.

Skolnick, J. H. (1966). *Justice without Trial: Law Enforcement in a Democratic Society*. New York: John Wiley and Sons.

———, and D. H. Bayley. (1988). *Community Policing: Issues and Practices around the World*. Washington, DC: National Institute of Justice.

Smith, A. H. (1899). *Village Life in China*. New York: Fleming H. Revell Co.

Smith, L. (1993). "The role of police in crime prevention." Paper presented to the Crime and Older People Conference. Adelaide: Australian Institute of Criminology.

Social and Economic Situation in Russian Federation (1994). Moscow: Gosstat.

Soloff, C. (n.d.). "Neighborhood Watch: An Information and Discussion Paper on Sustaining the Momentum." Working Paper. Adelaide, Australia: National Police Research Unit.

Sorgdrager, W. (1997). Changing Attitudes and Pragmatism. *European Journal on Criminal Policy and Research* 5:7-11.

South Australia, Ministry of Crime Prevention. (1990). "Together against Crime: Responses to the South Australian Crime Prevention Strategy: Conference Papers 1." Adelaide: South Australian Government.

Sparrow, M.K. (1993). *Information Systems and the Development of Policing*. Perspectives on Policing, No. 16. Washington, DC: National Institute of Justice.

Spiridonov, L., and Y. Gilinskiy. (1977). *Man as the Object of Sociological Research*. Leningrad: Leningrad University.

Sprenkel, V. D. (1962). *Legal Institutions in Manchu China*. London: University of London.

State of Crime in Russia—1997 (1998). Moscow: MVD RF.

Statistics Canada. (1992). *Policing in Canada*. Ottawa: Canadian Centre for Justice Statistics.

———. (1996). *1997 Canada Yearbook*. Ottawa: Minister of Industry.

Steinert, H. (1995). "The idea of prevention and the critique of instrumental reason." In Albrecht, G. and W. Ludwig-Mayerhofer (Eds). *Diversion and Informal Social Control*. Berlin: Walter de Gruyter and Co.

Stratta, E. (1992). "From policy to practice—Police community liaison in Britain." Unpublished paper from the Worcester, College of Higher Education.

Stuive, K., and E. Belt. (1997). Met twee benen in de wijk. *Het Tijdschrift voor de Politie* 4(April):4–7.

Sumner, C. (1994). *The Sociology of Deviance: An Obituary*. Buckingham: Open University Press.

Sun, Y. (1977). The reorganization of public security must be through the waging of a people's war. *Theory and Practice of Public Security* 6(3):1-4.

Sutton, A. (1991). "Crime prevention at the crossroads." Paper presented to the Victorian Good Neighbourhood State Workshop.

———. (1994). Community crime prevention: A national perspective. In D. Chappell and P. Wilson (Eds.), *The Australian Criminal Justice System: The Mid 1990s*. Sydney: Butterworths.

———, and J. Fisher. (1989). *Confronting Crime: The South Australian Crime Prevention Strategy*. Adelaide: Attorney-General's Department, Ministry of Crime Prevention.

Tamuno, T. N. (1970). *The Police in Modern Nigeria 1861–1956*. Ibadan, Nigeria: Ibadan University Press.

Terlouw, G. J. (1991). *Criminaliteitspreventie onder Allochtonen: Evaluatie van een Project voor Marokkaanse Jongeren.* Arnhem, Netherlands: Gouda Quint.

Tien, J. M., and T. F. Rich. (1994). The Hartford COMPASS program: Experiences with a weed and seed-related prpogram. In D. P. Rosenbaum (Ed.), *The Challenge of Community Policing: Testing the Promises*. Thousand Oaks, CA: Sage.

Topley, M. (1967). Chinese religion and rural cohesion in the nineteenth century. *Journal of the Hong Kong Branch of the Royal Asiatic Society* 9–44.

Trojanowicz, R. (1983). *An Evaluation of the Neighborhood Foot Patrol Program in Flint, Michigan*. East Lansing, MI: Michigan State University.

———. (1994). "The future of community policing." In D. P. Rosenbaum (Ed.), *The Challenge of Community Policing: Testing the Promises*. Thousand Oaks, CA: Sage.

———, and B. Bucqueroux. (1990). Community Policing: A Contemporary Perspective. Cincinnati: Anderson Publishing Co.

———, and M. Moore. (1988). The meaning of community in community policing. *Community Policing Series*, No. 15. East Lansing: Michigan State University Press, National Neighborhood Foot Patrol Center.

Troyer, R. J. (1989). Chinese social organization. In R. J. Troyer, J. P. Clark, and D. G. Rojek (Eds.), *Social Control in the PRC*. New York: Praeger.

———, J. P. Clark, and D. G. Rojek. (1989). *Social Control in the PRC*. New York: Praeger.

van der Vijver, C. D. (1998). Personal interview.
Van Dijk, J. (1990). Confronting crime: The Dutch experience. *Responses to the South Australian Crime Prevention Strategy*, Conference Series 1. Adelaide: South Australian Government.
———, and J. De Waard. (1991). A two-dimensional typology of crime prevention projects; with a bibliography. *Criminal Justice Abstracts* (September):483–503.
Van Erpicum, I. (1997). Jeugdcriminaliteit blijft zorgenkindje. *SEC* 11:5.
———. (1998). De evolutie van criminaliteitspreventie. *SEC* 12:22–24.
Van Oostveen, M. (1998). Europese uitwisseling van preventie. *SEC* 12:17.
Van Reenen, P. (1993). Het zwefende politiebestel. *Justitiële Verkenning* 19(4):7–36.
Veenbaas, R., and J. Noorda. (1997). Nieuwe prespectieven voor Amsterdamse jongeren. *SEC* 11:12–14.
Veno, A., and E. Veno. (1989). The police, riots and public order. In D. Chappell and P. Wilson (Eds.), *Australian Policing: Contemporary Issues*. North Ryde, Australia: Butterworths.
Versterking Jeugd Openbaar Ministerie. (1998). The Hague: Jeugdcriminaliteit, Ministerie van Justitie. http://www.minjust.nl/a_beleid/thema/jeugd/beleid/project/pr_11.htm
Vleesenbeck, V., A. Littooij, and J. Tuinder. (1996). Buurtagent en Communitarist. *Algemeem Politieblad* 13(22 June):4–7.
Vodinelic, V. (1985). *Criminalistics, Detection and Proving*. Skopje: Elisie Popovski-Marko Skopje.
Walker, S. (1998). *Policing in America: An Introduction*, 3rd ed. New York: McGraw-Hill.
———. (1992). *The Police in America: An Introduction*, 2nd ed. New York: McGraw-Hill.
Ward, C. M. (1997). Community crime prevention: Addressing background and foreground causes of criminal behavior. *Journal of Criminal Justice* 25:1–18.
Ward, R. (1985). *The police in China. Justice Quarterly* 2:111–115.
———. (1985a). The police of China. *Criminal Justice International* (Winter):7.
Weatheritt, M. (1986). *Innovation in Policing*. London: Croom Helm Ltd.
Weisburd, D., M. Amir, and O. Shalev. (1998). *Intermediate Research Report on the Evaluation of Community Policing Implementation in Israel*. Jerusalem: Ministry of Public Security.
Weisel, D. L., and J. E. Eck. (1994). Toward a practical approach to organizational change: Community policing initiatives in six cities. In D. P. Rosenbaum (Ed.), *The Challenge of Community Policing: Testing the Promises*. Thousand Oaks, CA: Sage.
Weisheit, R. A., L. E. Wells, and D. N. Falcone. (1994). Community policing in small town and rural America. *Crime and Delinquency* 40:549–567.
Wen, J. T. (1971). *Chinese Baiojia System*. Taipei, China: Shangwu yinshuguan.
Wen, X. H., and C. J. Chen. (1997). A brief discussion of social environment of close police and people's relationship. *Gongan Yanjiu* 4:25–27.
West, J. (1992). Problem-oriented policing: A team approach. In S. McKillop and J. Vernon (Eds.), *Proceedings of the National Overview on Crime Prevention Conference*. Adelaide: Australian Institute of Criminology.

Wilbur, M. C. (1978). Village government in China. *Journal of the Hong Kong Branch of the Royal Asiatic Society* 18:13–175.

Wilkinson, D. L. and D. P. Rosenbaum (1994). The effects of organizational structure on community policing: A comparison of two cities. In D.P. Rosenbaum (Ed.), *The Challenge of Community Policing: Testing the Promises*. Thousand Oaks, CA: Sage.

Williams, S. W. (1883). *The Middle Kingdom*, Vol. 1. New York: Scribner's.

Wilson, A. A., S. L. Greenblatt, and R. W. Wilson. (1977). *Deviance and Social Control in Chinese Society*. New York: Praeger.

Wilson, J.Q. (1968). *Varieties of Police Behavior*: Boston, MA: Harvard University Press.

Wilson, J. Q., and R. J. Herrnstein. (1985). *Crime & Human Nature*. New York: Touchstone Books.

———. and G. L. Kelling. (1982). Broken windows: The police and neighborhood safety. *The Atlantic Monthly* (March):29–38.

Wilson, L. (1986). Neighbourhood watch. *Legal Service Bulletin* 11:68–71.

Wittfogel, K. (1957). *Oriental Despotism: A Comparative Study of Total Power*. New Haven: Yale University Press.

Wong, A. K. (1971). Chinese voluntary associations in southeast Asian cities and the Kaifongs in Hong Kong. *Journal of the Hong Kong Branch of the Royal Asiatic Society* 11: 62–73.

Wong, K. C. (1994). Public security reform in China in the 1990s. In Brosseau and C. K. Lo (Eds.), *China Review*. Hong Kong: Chinese University of Hong Kong.

———. (1996). Police powers and control in the PRC: The history of *shoushen*. *Columbia Journal of Asian Law* 10:367–390.

———. (1996a). "The Origin of Communist Policing in China." Paper presented at a seminar at Universities Service Center, Chinese University of Hong Kong.

———. (1997). Sheltering for examination (*shoushen*) in the PRC: Law, policy, and practices. *Occasional Papers/Reprints Series in Contemporary Asian Studies*. College Park, MD: School of Law, University of Maryland.

———. (1997a). "Law of Assembly in the PRC and ROC: A Comparative Study of Police Powers." Paper presented at the Hong Kong Unification and China-Taiwan Relations Prospects Conference, Chinese University of Hong Kong.

———. (1998). A reflection on police abuse of powers in the PRC. *Police Quarterly* 1:89–112.

———. (1998a, forthcoming). Black's theory on the behavior of law revisited III: Law as more or less governmental social control. *International Journal of the Sociology of Law*.

Wright, K. N. (1980). The desirability of goal conflict within the criminal justice system. *Journal of Criminal Justice* 9:19–31.

Wu, S. C. (1995). *Traditional Chinese Legal Culture*. Beijing: Beijing daixue chubanshe.

Wycoff, M. A. (1995). Community policing strategies. *NIJ Research Review*. Washington, DC: U.S. Department of Justice.

Xiao, X. W. (1997). Moli Ba Shu Zhi Jian. *Renmin Jingcha* 16:15.

Yang, M. C. (1947). *A Chinese Village: Taitou Shantung Province.* New York: Columbia University Press.

Yang, S. L., and L. Fang (1987). *Historical Documents of Shan-Gan-Ning Border Area.* Beijing: Falu chubanshe.

Yanli, L. (1994). *A Comprehensive Public Security Post Business Manual.* Beijing: Zhongguo renmin gongan daixue chubanshe.

Zhang, J. (1997). Analysis of the public order situation, background, and countermeasures during the social transformation period. *Theory and Practice of Public Security* 6(3):4–12.

Zhang, L. N., D. K. Zhou, S. F. Messner, A. E. Liska, M. D. Krohn, J. H. Liu, and Z. Lu. (1996). Crime prevention in a communiarian society: Bang-Jiao and Tiao-Jie in the PRC. *Justice Quarterly* 13:199–222.

Zhang, X. B., and Y. L. Han. (1987). *Legal History of Chinese Revolution.* Beijing: Zhongguo shehui kexue chubanshe.

Zhongguo Gongan Bake Quanshu Editorial Committee. (1989). *Chinese Public Security Encyclopedia.* Jilin: Jilin chubanshe.

INDEX

D

E

I

J

K

T

U

V

W

Y